Index

Abraham, 105, 108
Acceleration, 20
Alexander, S., 141, 152-53, 155, 158-60, 169
Anthropic Principle, 67-68, 72
Anatman, 87
Apocalyptic, 114
Aranyakas, 76
Arjuna, 77-78
Asamkhyeya, 89
Ashoka, 77
Atman, 78, 87
Atom, size of, 25
Avatamsaka school, 91, 160-61

Bhagavad Gita, 77-78, 80
Big Bang, 37, 39, 50, 68-72, 136-37, 145; Theory, 55-59, 138, 162-63
Black body radiation, 26, 58
Black holes, 63, 66-67. See also Heisenberg, W.
Bodhisattva, 89
Bohr, N., 25, 36; Complementarity Principle, 6, 29, 167-68
Boltzmann, L., 44-45, 52
Brahma, 80-81, 141
Brahman, 76-84, 100, 137, 140-42, 146-47, 171; Nirguna, 80; Saguna, 80
Brahmanas, 76
Buber, M., 117, 124-27, 130; I-It concept, 126-27, 165-66; I-Thou concept, 125-27, 148, 154-55, 165-66

Buddha, Gautama, 85-87, 89
Buddhism, 77, 85-86, 94-95; cosmology, 89-92, 140; sects, 87-88; view of time, 87-93, 147-48, 159-61

Canaan, 105-7; Canaanite calendar, 106; myths, 110-11
Carter, B., 68
Christ, Jesus, 109; ministry, 109, 111; resurrection, 109, 111; second coming, 110, 140, 153-54
Complementarity Principle, *See* Bohr, N.
Conditioned, definition of, 86; Conditioned Coproduction (Dependent Origination), 86, 93
Confucianism, 94-95
Conze, E., 86
Copernican Principle, 67
Creation myths, 104, 136; Canaanite (Ugarit), 111; Enuma Elish, 111
Cullman, O., 111-12

Daniel, 114-16, 139
Davisson and Germer, 28
deBroglie, L., 27-28, 31, 33
Denbigh, K., 2, 166-67, 174
Determinism, 24, 40, 53
Dharma, 81
Diffraction, 30
Dirac, P.A.M., 65

191

Dodd, C. H., 113
Doppler effect, 56; blue shift, 66; red shift, 56
Dunne, J. W., 169
Dyson, F. J., 65-66, 140

Eddington, Sir A. 6, 48, 150, 174
Eight-Fold Path, 85-87
Einstein, A., 10-14, 20-23, 25, 62; formula $E = mc^2$, 19, 38, 63, 159; Nobel prize, 24, 26; photoelectric effect, 26
Electron volt, 59, 63; definition, 59; GeV, definition, 59
Elementary particle physics, 59-61; definition, 55
Eliade, M., 1, 103-5, 108-9, 144, 172; sacred and profane time, 103, 165
Eliot, C., 160
Eliot, T. S., 171
Entropy, 44-49, 52-53
Equivalence, Principle of, 20
Eschatology, 113, 153-54
Ether, 11-13
Exodus, 105-6

Four Noble Truths, 85
Fraser, J. T., 22-23, 149, 170-71, 174

Gamow, G., 58
Genesis, 56, 110-12, 136-38, 149
Govinda, Lama A., 92, 148, 157, 163
Grand Unified Theories, GUT, 61, 64, 71, 145; Super-GUT or Supergravity, 61, 170
Gravitation, 20-21, 61

Hawking, S. W., 65; Collins and, 68
Heisenberg, W., 29, 31, 36; matrix mechanics, 31; relation to Black Holes, 65; relation to complementarity, 33; Uncertainty Principle, 29-31, 38, 47-48, 70,147, 159
Hinayana, 88
Hinduism, 75-79, 81, 171; cosmology, 79-82, 134-35, 146-47; view of time, 82-84
Husserl, E., 151, 154

I Ching (Classic of changes), 96, 155; hexagrams, 96
I-Hsing, 97, 142
Isaac, 105, 108
Israelites, 105-10, 116
Isvara, 80
I-Thou. *See* Buber, M.

Jainism, 77
Joshua, 106
Jastrow, R., 66

Kairos, 107
Kalpa, 81, 89
Karma, 78, 81
K-meson, 50, 54
Krishna, 77-78
Kronos, 6

Lao-tzu, 97-98, 135
Layzer, D., 49
Lee, T. D., 39, 161-62
Light, 21, 26-27, 149; cone, 17-19; velocity of, 11-17, 56
Long, C. H., 136
Longfellow. W. W., 172

Macroscopic world, 25, 44
Madhva, 79, 82
Madhyamikas, 88
Magnetic monopoles, 58-59
Mahabharata, 77-78
Mahayana, 88-93
Marsh, J., 112-13

Mass, 19, 159
Maya, 80-81
Mayan calendar, 142
Mercury, precession of, 21
Merleau-Ponty, M., 151, 154
Messiah, 110, 112, 153
Messianic Age, 112, 114, 154
Michelson and Morley, 12-13
Microscopic world, 25, 44
Minkowski, H., 14-15
Moksha, 81, 140, 144
Moses, 105-6

Nagarjuna, 88
Nagasena, 90
Needham, J., 96, 100
New Testament, 114, 135, 139-40
Newton, Isaac, 2, 10
Nirvana, 86-87, 89-92, 140
Nucleus, size of, 25
Nyaya school, 79, 82

Old Testament, 108-15, 136-40

Pannikar, R., 144, 174
Park, D., 2, 43, 166-67, 170, 172
Penzias and Wilson, 57, 60
Philosophy, of becoming, 22, 61-62; of being, 22, 61-62
Photoelecric effect. *See* Einstein, A.
Planck, M., 26, 28, 30-31
Planck's constant, 26
Plato, 144-45
Prakriti, 80
Prigogine, I., 51-53, 54
Pudgalavadins, 88
Puranas, 78, 80
Pureland sect, 89, 95
Purva-Mimamsa, 79, 82

"Quantum foam." *See* Wheeler, J. A.
Quantum Mechanics, 31-33, 34

Quantum Theory, 24, 34-35; Observer Theory, 35-37

Ramanuja, 79, 82
Ramayana, 78
Red Giant, 64, 139
Relativity, Theory of, 12, 158-9; general, 20-23; Principle of, 11; special, 12-19; transformation equations, 14-15, 159
Revelation, 114-16, 139-40
Riemann, G. F. D., 20
Robinson, R. H., 85

St. Augustine, 56, 113, 117-20, 136-38, 145, 174; concept of time, 2, 118-20, 130, 162; God's "eternal now," 148
Sakharov, A., 49, 142, 170
Samsara, 77, 86, 92
Sankhya school, 79
Sarvastivadin, 88
Schlegel, R., 48, 147
Schramm, D. N., 62, 72
Schroedinger, E., 31-32, 46, 148
Shankara, 79, 82-83
Shiva, 80
Simultaneity, 16-17
Skandhas, 87-88
Space, contraction, 16, 19; curvature, 20; Interdependence with time, 14-16; quantization, 57-59, 161-62
"Spectrum of Intellectual Pursuits," 5, 74
Spinoza, 24
Sutras, 79
Suzuki, D. T., 91, 157, 161
Supergravity. *See* Grand Unified Theories

Tantraism, 91
Tao te Ching, 98-100, 135
Taoism, 95-100, 171; cosmology,

99; Tao, 98-100, 137, 142, 146-47, 171
Teilhard de Chardin, 117-18, 127-30, 140-41, 152-53, 155; Omega Point, 129-31, 141, 153
Thermodynamics, 44; First Law of, 46; Second Law of, 46-48
Time, arrow of, 48-51, 150-56; biblical, 107-12, 153-55; cosmologic arrow, 50-51, 53, 57; cyclic, 49, 67, 102-5, 142, 146-47; dilation, 16, 19; historical arrow, 49-51, 53; interdependence with space, 14-16, 157-60; quantization, 37-39, 161-62; reversible, 42-44, 51-52, 147-48; thermodynamic arrow, 48-51, 53; timelessness, 83-84, 144-49
Transformation equation, *See* Relativity.
Trefil, J., 61, 137
Twin paradox, 21

Uncertainty Principle. *See* Heisenberg, W.
Universe, age of, 47, 59, 72; cyclic, 49, 67, 142, 146-47; inflationary, 59; open or closed, 62-67
Upanishads, 76, 80; *Chandogya*, 81; *Svetasvatara*, 77, 80

Vaisesika school, 79, 82

Vedanta, 79, 82-83
Vedas, 75; *Atharva*, 76, 83-84; *Rig*, 75, 81
Vishnu, 80
Von Franz, M., 96
Von Rad, G., 108
Von Schiller, 177

Wave-particle duality, 27-29, 168
Welch, H., 100
Wheeler, J. A., 36-39, 63, 68-72, 136-37, 145; concept of time, 70, 147, 162-63; Law of Mutability, 70, 136; matter and space, 21, 57; "quantum foam," 38, 62, 142, 147, 170
Whitehead, A. N., 117, 120-24, 130, 140-41, 172; Concept of time, 123-24, 152-53, 155, 157-60, 162; religious views, 122-23
Wu Ching (5 Classics), 95
Wu-wei, 98

Yahweh, 105-9, 153
Yin-yang, 96, 167-68; symbol, 97
Yoga, Buddhist, 90, 148; Hindu school, 79; yogin (practitioner), 90
Yogacara, 88
Yuga, 80-81

Zurvan, 5

About the Author

Lawrence Fagg is Research Professor in Nuclear Physics at the Catholic University of America in Washington, D.C. and a Fellow of the American Physical Society. In addition to a Ph.D. in Nuclear Physics from John Hopkins University, he holds a Master's degree in Religion from George Washington University.

A vice president of the Institute on Religion in an Age of Science and a member of the International Society for the Study of Time, Dr. Fagg has lectured and presented seminars in the area of science and religion for the Forum for the Humanities of the Washington School of Psychiatry and also for the Institute on Religion in an Age of Science. He has lectured on the subject of time in physics and religion at meetings of the George Mason University and the Catholic University of America. He has also lectured on his work in physics, which concerns electron scattering from nuclei, at universities, laboratories, and conferences in the United States, Canada, Europe, Japan, and Australia.

QUEST BOOKS
are published by
The Theosophical Society in America,
a branch of a world organization
dedicated to the promotion of brotherhood and
the encouragement of the study of religion,
philosophy, and science, to the end that man may
better understand himself and his place in
the universe. The Society stands for complete
freedom of individual search and belief.
In the Theosophical Classics Series
well-known occult works are made
available in popular editions.

Additional Quest books are available on a wide spectrum of subject matter such as transpersonal psychology, philosophy, comparative religion, occultism, meditation, etc. Write for our free catalog.

The Theosophical Publishing House
306 West Geneva Road
Wheaton, Illinois 60189

Two Faces of Time

Cover art by Jane A. Evans

Two Faces of Time

BY LAWRENCE W. FAGG

*This publication made possible with
the assistance of the Kern Foundation*

 The Theosophical Publishing House
Wheaton, Ill. U.S.A.
Madras, India / London, England

• Copyright Lawrence W. Fagg, 1985.
A Quest original. First edition, 1985.

All Rights reserved. No part of this book may be reproduced in any manner without written permission except for quotations embodied in critical articles or reviews. For additional information write to:

The Theosophical Publishing House
306 West Geneva Road
Wheaton, Illinois 60189

A publication of the Theosophical Publishing House, a department of the Theosophical Society in America.

Library of Congress Cataloging in Publication Data

Fagg, Lawrence W., 1923-
  Two faces of time

  (A Quest book)
  "A Quest original"—P.
  Bibliography: p.
  Includes index.
  1. Time I. Title
BD638.F34   1985   115   85-40412
ISBN 0-8356-0599-X (pbk.)

Printed in the United States of America

Contents

Preface, vii

Introduction: The Mystery of Time, 1

I TIME IN THE PHYSICAL WORLD

    1 Relativistic Time, 9

    2 Time in the Quantum Theory, 24

    3 Reversible and Irreversible Time, 42

    4 Focus on the Universe: Cosmology and Elementary Particle Physics, 55

II SOME RELIGIOUS VIEWS OF TIME

    5 Hindu Cosmic Cycles: Manifestations of the Timeless Brahman, 75

    6 Buddhism, Nirvana, and Time, 85

    7 China and Taoism, 94

    8 From Cyclical Rituals to Judeo-Christian Linearity, 102

    9 Four Western Theologians, 117

III PHYSICAL AND RELIGIOUS TIME CONCEPTS COMPARED

    10 Beginnings, Endings, Cycles, Durations, 135

    11 Timelessness, 144

    12 The Arrow of Time, 150

    13 Interdependence of Time and Space, 157

    14 Time: A Duality or Unity? 164

Notes, 179

Index, 191

Preface

In our quietest, most thoughtful moments, we can watch the second hand of a clock precisely tick off the seconds and at the same time, resting in the living present, sense one moment gently and indistinguishably merge into the next. Time is an enigma. Trying to grasp and look at it is like trying to grasp water in one's hand.

For this reason alone it is a fascinating subject which has engaged thinkers for millennia. The literature about the manifold aspects of time, of course, is vast. However, it struck me several years ago that one approach in attempting to narrow the subject and concurrently to enhance our understanding of time would be to examine it from what many consider the extreme perspectives of physics, the "hardest of sciences," and religion, the "softest of humanities."

The pursuit of such an approach has proven to be not only enlightening as far as knowledge is concerned but also surprisingly stimulating to the imagination. One has the opportunity to vicariously sense the infectious curiosity of the physicist exploring on the one hand the "world of the very small": atoms, nuclei, and quarks; and on the other, the "world of the very large": stars, galaxies, and black holes. When this is mixed with the power and beauty of the profound and incisive spiritual insights found in a spectrum of religious traditions, the result is a learning experience rich in diversity and imaginative reflection.

However, the examination of time from the two perspectives, physics and religion, still involves a very extensive amount of source material. This has meant that only carefully chosen and representative highlights could be presented. Unfortunately, for example, I felt forced to leave out time concepts from such religious traditions as Islam, Zoroastrianism, and Jainism.

Furthermore, this book is written as an introduction to selected concepts of time for the average college-educated lay reader without training in either physics or religion. Consequently, I felt it imperative to devote considerable space to background material in many of the areas of modern physics as well as in the selected religions in order to render the time concepts more understandable and meaningful. In particular, I have tried to present the necessary principles of physics in as descriptive and conceptual a fashion as possible with an absolute minimum of mathematics.

On a subject as elusive as time, firm answers to questions are hard to come by. Often the best that can be done is to frame intelligent questions. One of the prime purposes of this book is to furnish the layman with enough introductory background to reflect meaningfully on such questions. Another purpose is to stimulate creative speculation and to instill a sense of wonder. This is not a book of deep scholastic study or rigorous philosophic investigation. Philosophic concepts have only been included occasionally when it was deemed helpful in enhancing the understanding of the physical or religious views under discussion. Therefore, this is a book of basic information, of questioning, and of speculation and wonder, which may serve to open the door to the pursuit of further interest.

The material is organized in three parts. The first deals with the time concepts that can be drawn from relativity, quantum theory, thermodynamics, cosmology, and elementary particle physics. The next part treats selected views of time as found in the Hindu, Buddhist, Taoist, and Judeo-Christian traditions. At the end of each chapter in the first two parts a summary of the time-related points treated in the chapter is presented. It is largely these points that are considered in the chapters of Part III. In this part comparisons are made between the physical and religious perceptions, primarily with respect to at least six aspects of time: 1) the beginning of time, 2) the cosmological duration of time, 3) the possible end of time, 4) the notion of timelessness, 5) the irreversibility of time, and 6) the interrelatedness of time and space. Cyclical time also plays a significant role in the first three aspects. Finally, I conclude with a discussion of whether the two views of time constitute a duality or can be unified in some way.

I am very grateful to Prof. Robert John Russell, Dr. Carol Norman, and Mr. John Lewis for prereading the manuscript. I also greatly appreciate helpful discussions with Prof. John Wheeler, Dr. J. T. Fraser, Prof. David Park, Prof. Harry Yeide, Prof. Robert John Russell, Prof. Robert G. Jones, Prof. Alf Hiltebeitel, Dr. Anandita Balslev, and Ms. Carolyn Elkins. I particulary wish to express my gratitude to Prof. Yeide for having guided me into such an intriguing field during my graduate study in religion. The patience, support, and forbearance of Mrs. Kay Winner in the typing and word processing of the manuscript at The Catholic University of America Computer Center were absolutely invaluable for the production of this book.

INTRODUCTION

The Mystery of Time

Time is a mystery. Time along with notions of timelessness has probably fascinated and engrossed men and women since evolution first endowed them with consciousness. Certainly from the very beginning of recorded history an awareness of time has played a central role in our cultural development. The witnessing of the fundamental phenomena of before and after, birth and death, and the cycles of sun and moon and the seasons has always engendered awe and wonder. Throughout history events such as these and the notions of time they imply have evoked responses about the nature of time. These responses have been as varied as mankind's intuitive depth and intellectual limits. They have ranged from aboriginal religious rituals worshipping the sun to meticulous mathematical and philosophical analyses of the characteristics of time. Human thought about time has had a long history with a broad spectrum of theories, concepts, and insights.

However, if we examine this history up to the 20th century, there is considerable evidence to support the claim that time concepts can be meaningfully classified into two general categories: religious and scientific. The religious concepts of time usually, but not always, tend to arise from subjective experience, intuition, revelation, and spiritual insight. The scientific concepts of time usually, but not always, tend to result from objective observation, quantitative measurement, mathematical or physical analysis, and logical, rational thinking.

The classification of time into two or more categories approximately correlating with the above is not new. The renowned religious scholar, Mircea Eliade, speaks of sacred and profane time.[1] The former is religious time in the sense that it is the time each year when the archaic worshipper relived the mythical moment of creation or some great triumph of a mythical hero.

The latter for Eliade is the time in between, lived in life's day-to-day exigencies. While, by modern-day standards this kind of time is not strictly scientific, it is the time that might be made so by objective observation.

More recently physicist David Park has spoken of two times he calls Time 1 and Time 2.[2] The first is the time of physical theory and objective measurements, the second essentially the time of human consciousness and subjective experience. In a recent study by scientist Kenneth Denbigh he speaks of three concepts of time; 1) that of conscious awareness, 2) that of theoretical physics, and 3) that of thermodynamics and biology.[3] In terms of Park's two broader categories, the last two of Denbigh's concepts could perhaps be included in a general objective concept of time.

These are but a few examples of a considerable body of opinion which sorts time concepts into at least two broad categories, one of a generally subjective character and the other generally objective. So, then, what justifies a division into scientific and religious categories? As implied earlier, at least until modern times a preponderance of mankind's recorded subjective notions about time has been in a religious context, while most of its objective views of time have been in a logically philosophical or scientific context. There have, of course, been some exceptions to such a general statement. For example, some of St. Augustine's thought concerning time had exquisitely analytical and objective aspects, and Newton was strongly influenced by some distinctly religious viewpoints. Nevertheless, there is considerable validity to the supposition that throughout most of our cultural history, religious and scientific concepts of time have generally followed parallel but basically separate courses.

However, toward the end of the 19th century with the beginning of the "scientific revolution," the interaction between science and religion intensified. Let us very briefly summarize this interaction, which started roughly a century ago when Darwin came forward with his theory of evolution which profoundly altered our idea of who we are. Freud and Marx had set down ideas and systematics which approached "scientizing" the fields of psychology and politics. However, the dramatic climax of the early stages of the revolution came at the beginning of this century with the advent of the theory of relativity and the quantum theory, which effected a prodigious breakthrough in our conception of the physical world.

All of this made for an enormous burst of rational, objective, self-consistent knowledge and thought about who we are and what the world is around us. This new knowledge flew in the face of many previously held ideas that were religious in origin. Thus the scientific revolution was often seen as a devastating onslaught on the integrity of religious thought. Religion seemed to be in retreat.

On the other hand, the scientific revolution was regarded by many thinkers as opening up a new vista of opportunity to reexamine and recompare scientific and religious concepts. In part because of this, but mostly because of the strength of its doctrines and the core of faithful believers, religion has continued as a significant influence in modern culture. Indeed, with the growing disillusion with science in the last two decades, brought about by antinuclear and pro-environmental sentiment, there has been an attendant and noticeable increase in the attention given to the nonrational and spiritual part of our nature. Some of this growth may also be an indirect byproduct of the expanded influence that the fields of psychology and psychiatry have exercised since World War II. Whatever its cause there has been, especially in the last two decades, an impressive increase in the number of religious and quasireligious cults, of interpersonal sensitivity and spiritual development enterprises, and of studies of varying quality and validity in parapsychology and extrasensory perception.

Thus far there seems to be no clear distillation of what all this religious and psychological activity has produced with respect to religious conceptions of the world, not to mention relationships between these conceptions and those of science. However, one can sense, especially in recent years, an underlying restlessness in both the scientific and religious communities; both seem engaged in an inarticulate groping for sweeping and fundamental reformulations of the scientific and spiritual conceptions of the universe, its content, and workings.

Some of this probing has taken the form of attempts to find

similarities, analogies, or correlations in scientific (especially physical) and religious conceptual structures. There is already evidence for this to be found not only in several recent books for popular consumption,[4] but also in books at a deeper and more scholarly level.[5] Furthermore, there are now active organizations whose specific mission is the search for meaningful relationships between science and religion.* In addition, many colleges and universities have been offering courses in science and religion for years.

All of this activity, in my opinion, comes from a realization that religion needs science to revitalize its theological structure and to refine its moral and spiritual values for more effective meaning in the 20th century. On the other hand, science may need religion for more enlightened and altruistic motivation, for deeper, more powerful intuitive insights into nature and its relation to mankind, and for enrichment of its meaning in human culture.

The sense of integration that a closer interaction between science and religion suggests has yet a long way to develop. Despite the relatively recent interest in the spiritual and intuitive aspects of our character mentioned earlier, as products of the scientific revolution we are still inundated in rationality, analysis, and objectivity. Computers and a vast variety of electronic devices from televisions to stereos occupy large parts of our attention. Friends and relatives are now objectified and called "support systems." For many of us the intuitive, irrational side of our natures seems to be near total atrophy.

This was not so in earlier days. Anthropologists tell us that the human brain, after having experienced an incredible evolutionary growth of ten cubic inches every hundred thousand years, has not changed significantly in the last hundred thousand. This means that although men and women of five thousand years ago did not have access to as full a flowering of scientific sophistication as we enjoy today, they may have been just as intelligent. What they lacked in exquisite scientific ability, they may have

---

*For example, the Institute on Religion in an Age of Science, which sponsors *Zygon*, a journal of science and religion; the Center for Theology and Natural Sciences, Berkeley, CA; the Isthmus Institute, Dallas, TX.

more than made up for in a refined capacity for intuition and spiritual insight. This in turn means that the literature that has been uncovered which was based on ancient insight should be looked upon as having far more than historical value. Such literature constitutes a rich repository of the accumulated wisdom of centuries of thought and spiritual contemplation. It is there, a treasured heritage, waiting for us, challenging us to touch, respond, and relate, with all of the intuitive perception and spiritual sensitivity we can muster.

In this book I examine at an introductory level some relationships with respect to time between one particular science—physics—and certain religious traditions. In this century the greatest changes in our notions of time and space have come from physics. These arose not only from the vast changes in our view of the world due to relativity and quantum theories, but also from great advances in the fields of thermodynamics, astronomy, cosmology, and elementary particles.

Physics can be considered to provide a more precisely focused perspective about time because it is at the opposite extreme from religion in what might be termed the "spectrum of intellectual pursuits." Such a spectrum would range from studies of the most refined precision and detail to more and more holistic studies embracing ever larger structures of human concern. More specifically, physics, with its study of subnuclear particles, nuclei, and atoms, would stand at one end of the spectrum, followed by chemistry with its molecules, and biology with its cells and organisms; hence through zoology, anthropology, psychology, sociology; then into the humanities, with philosophy, literature, and finally religion. At this far end of the spectrum I will focus on selected views from major religious traditions, presenting important contrasts and correlations with views of physical time.

Underlying attempts at seeking relationships between such disparate points of view on the subject of time, or indeed on any other subject that can be meaningfully so addressed, is a struggle to achieve some kind of unity and deeper simplicity. In ancient times this striving often found expression by deifying time as an anthropomorphic god. In a well-known and ancient West Iranian (non-Zarathustran) myth Zurvan, father of Ahura Mazda, is Time Unbounded. The father of the Greek god, Zeus, was Kronos, who many early thinkers identified as the deified personification of Chronos, Time.

The foregoing remarks have implied that another purpose of this book is to explore possibilities of a unified concept of time. In such a concept, the religious (or subjective and intuitive) and the scientific (or objective and rational) views might be, if not reduced to the same thing, at least seen as different ways of looking at a common construct. This suggests the possible use of a generalization of Bohr's Complementarity Principle (see Chapter 14), by regarding the two views of time as complementary and constituting two mutually exclusive yet equally valid expressions of time, both of which are necessary for a complete picture.[6]

In any case the question as to whether the two views of time can be unified is still an open one. Such a question impacts powerfully on the deeper question of the centrality and meaning of time as some kind of linking mechanism for the disparate aspects of our experience, and possibly of all activity in the universe. Perhaps this notion, in relation to the physical and religious concepts addressed in this book, was best expressed by physicist and natural philosopher Sir Arthur Eddington, who once observed: "In any attempt to bridge the domains of experience belonging to the spiritual and physical sides of our nature, time occupies the key position."[7]

I

Time in the Physical World

In this Part some of the elementary concepts of the relativity and quantum theories are presented, as well as concepts from thermodynamics, cosmology, and elementary particle physics. Time-related aspects of these areas of physics are highlighted and discussed, with particular emphasis being placed on those aspects of time that will be compared with the religious concepts treated in Part II.

1

Relativistic Time

The enormous revolution that occurred in physical science at the beginning of this century can hardly be overdramatized. Within a few years of each other, starting with the very first year of this century, two theories—the quantum theory and the theory of relativity—disintegrated the foundations of classical physics, and opened up conceptual vistas previously undreamed of. These two theories have served us ever since as the pillars of modern physical thought and have thus far remained essentially unchallenged by all experimental tests. In particular, each of these theories has had its own unique impact in altering our concepts of time and causality, even changing the meaning of time in our own lives.

In this chapter I will outline some of the essential features of the theory of relativity. It is not at all intended to be a complete and rigorous discourse on the subject, since there are many texts available at all levels of complexity and detail.[1] What is intended is sufficient descriptive and conceptual background for the nonscientific reader to understand the aspects of the theory relating to the time concepts that I wish to emphasize and discuss. Some of these aspects will receive further elaboration with the presentation of some philosophic implications given by selected thinkers on the subject, which may help us later in making comparisons with religious time concepts.

The Special Theory of Relativity

For centuries in Western civilization classic physical philosophers had generally considered time as a totally separate entity, completely divorced from space and the rest of the universe. It flowed linearly and endlessly on in its unique, independent way. This view was brought to its clearest expression with Isaac Newton who stated: "Absolute, True, Mathematical time, of itself, and from its own nature flows equably without regard to anything external . . . ."[2] In particular, absolute time was considered entirely separate from absolute space, in which there was a strong belief also.*

This lofty, heretofore unassailable, autonomy of time was shattered by Einstein's theoretical work. Unlike others, Einstein did not try to preserve the existing theory with a patchwork of ad hoc supports contrived to explain some of the inconsistencies made apparent by experimental results. Instead he devised a new theory with a new conceptual basis.

How did such revolutionary thought spring from such a one as Einstein? In school he was in many respects a maverick, who often aroused the displeasure of his professors because of his independent interests and thinking. As a result, on graduation he could not find a satisfactory academic position and finally took a job in the Swiss patent office in Bern.

Although this may at first seem to be an unlikely atmosphere for the generation of anything like a theory of relativity, it proved to be a most auspicious environment. This was in part because the job itself in a fundamental way helped train him and complemented his after-hours efforts in developing relativity. His boss was a kind and wise man, but a strict one. He taught Einstein to see through to the core concept in any patent he was examining and to express succinctly why it would or would not work.

Einstein turned this seven years of work and training to good use by, among other things, questioning the long-accepted Newtonian view of time as an independent, transcendent quantity. It was his insight of time as interdependent and interactive in the

---

*This belief in absolute space was founded on the notion that either there existed somewhere a completely stationary place or a stationary and pervasive medium with respect to which all motion of material bodies (a baseball, a train, the earth, the sun, etc.) could be referred; i.e., somewhere or somehow in the universe there had to be a fixed spatial anchor.

physical world that was a major step in the creation of his new theory.

For physicists of that day perhaps the most significant and fundamental inconsistency, ultimately indicating the need for a new theory, arose from the general belief in the late 19th century that there existed an "ether" which filled the absolute space and was essentially identified with it. Since it had been experimentally well established that light exhibited a wavelike character, it was naturally supposed that a universal, stationary medium, the ether, must exist in which these waves propagated. The common experience of pressure waves propagated in water and sound waves in air were undoubtedly among the analogies upon which the supposition was based.

With the supposition of absolute space (ether) and absolute time, classical mechanics had managed to remain viable through the use of what was known as the "principle of relativity" (not to be confused with Einstein's theory). In essence the principle stated that the laws of mechanics should be independent of whether we made measurements when stationary in the so-called absolute ether or when moving at some constant velocity relative to it. Indeed the principle should hold for any observer moving at some fixed velocity relative to another. For example, if you throw a ball vertically in the air, it returns, and you catch it; the motion is described in the same way by the laws of mechanics whether you are standing on the ground or in a train moving at constant velocity.

While the "principle of relativity" was thought to hold for mechanical phenomena, it was considered not to be valid for light and other electromagnetic phenomena. For example, assuming the principle holds and the absolute ether exists, measurements of the velocity of light (which is electromagnetic radiation) were expected to vary depending on the frame of reference in which the measurement was made. Here frame of reference essentially means the observer's local surroundings with respect to which he is stationary, i.e. the earth, a train, or a spaceship. The measurement then would be different depending on whether it was made in absolute space, the earth, or a spaceship. For example, a spaceship observer measuring the velocity of solar light would record a different value depending on whether the ship was moving toward, or away from, the sun. Thus, the true velocity of light should only be observable relative to the absolute and stationary ether, and would appear to have a different velocity with respect to, say, a space shuttle moving relative to the ether.

The crucial experiment designed to show that the principle of relativity was not applicable to light as well as to confirm belief in the existence of the ether was performed by Michelson and Morley in 1887.[3] If the ether were really present, then their apparatus situated on the earth's surface must be traveling through this medium at a certain velocity by virtue of the fact that the earth is rotating as well as orbiting around the sun. This being the case, mathematical predictions based on the principle of relativity indicated that there should be a very small but detectable difference between the transit time of light traveling a path back and forth along this direction of the earth's motion and the time for transit of an equal path transverse to this direction. The experiment is depicted in Figure 1. Therefore if the principle applied and the ether existed, the apparatus should have been able to detect such a difference. It did not.

The effect of the result of this experiment was extremely profound, for, in one stroke, it demolished the time-honored belief in absolute space, an absolute reference medium or ether, relative to which all velocities can be measured. Einstein used this result to develop a totally relativized principle of relativity which stated that there was no such absolute frame of reference. It is interesting to note that he never actually claimed that the absolute ether did not exist, only that, whether it existed or not, it could not be detected. In any case, this meant that all frames were on an equal footing, and all velocitites are relative. In essence this was the *first postulate* of his *special theory of relativity* by which he freed the earlier "principle of relativity" from its anchor to the notion of an absolute time and space. He "democratized" the principle.

That was not all. Einstein also showed that his new principle of relativity could be applied not only to mechanical but also to electromagnetic phenomena (light) through the use of his *second postulate*. It states that all measurements of the velocity of light give the same result regardless of the frame of reference in which

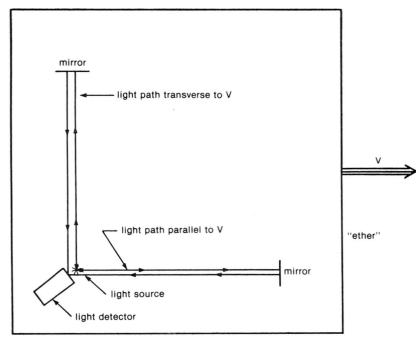

Figure 1. Illustration of the principle *of the Michelson-Morley Experiment. By virtue of the earth's motion, the apparatus in the square is moving at a velocity V with respect to the "ether." The actual apparatus was considerably more complicated.*

they are made, e.g. regardless of whether made on the earth or a spaceship.* This may at first seem strange and inconsistent with our ideas of motion in daily life, but it is at the very heart of Einstein's theory and it is what makes it truly relativistic.

With these two postulates as a foundation, a tremendous unification was possible. The null result of the Michelson-Morley experiment was explained, and both mechanical and electromagnetic phenomena could be embraced in a common theoretical framework. The beautiful simplicity with which the contradictions in physical theory had been resolved came from a man who

---

*The velocity of light is about 186,000 miles per second; it is the limiting velocity, faster than which nothing has been observed to move.

loved simplicity so much he often forgot or refused to wear socks at formal occasions. A little thought may impress us with the reasonableness of Einstein's postulates, because it would be a strange, unordered, and complicated universe indeed if the results of a physical observation depended on where or when it was performed.

This independence of a particular location, velocity, and time necessitated a certain symmetry in the equations mathematically describing the phenomena. In particular, it required deriving mathematical equations which would relate the measurements made in one frame of reference to those in another, moving at some velocity relative to the first, in such a way as to conform to the postulates. Such equations are known as transformation equations;* that is, they transform or relate the space and time measurements in one frame of reference to those in another.

A unique feature embodied in Einstein's theory, and particularly in the transformation equations, was that time was now placed on an essentially equal footing with space. To the three spatial dimensions a fourth, time, was added, so that time and the three space dimensions are now combined into a unified four-dimensional system or "space-time." In order to put time into the same terms as the three spatial dimensions, it must be multiplied by the velocity of light. This is similar to your driving a car at fifty miles an hour and saying, "I have two hours of driving ahead," but realizing that you can "spatialize" the statement by multiplying the time by your speed, and saying, "I have one hundred miles of driving ahead." The velocity of light then serves as a link between time and space. This four-dimensional system actually was first described by Minkowski and was also adapted by Einstein for his theory.

For us the most significant result of time being incorporated as a fourth dimension in the relativistic transformation equations is that time and space become inextricably interdependent, or, in a sense, mixed. Let us see in more detail how this is so.

---

*Actually Einstein found it was possible to adapt the transformation equations used in electrodynamics and already formulated by Lorentz. These he showed could be generalized for application to all physical phenomena—mechanical, electromagnetic, nuclear, etc.

Suppose there is an observer named Andrea in a frame of reference moving at some high, but constant, velocity with respect to Bob, an observer in another frame, and each observer is recording some physical event. An example would be Andrea in a spaceship (her reference frame) moving with respect to Bob in his laboratory on Earth, the respective observers studying a solar burst or flare. In relating her measurements on the event to those made on the event by Bob, Andrea will find that her space (distance) measurements in general not only depend on the space measurements in Bob's laboratory, but also on time measurement there. She will further find that her time measurement will not only depend on the time in Bob's laboratory, but also in general on space measurements there.* As will be discussed later, these effects are only noticeable when the relative velocity of Andrea and Bob is very high; they become more apparent as the relative velocity approaches that of light.

If Bob wished to determine his measurements in terms of Andrea's, he would make observations from an entirely reciprocal viewpoint and would find himself in a similar situation to Andrea. His time measurement would in general depend on both time and space measurements in Andrea's spaceship, etc. Thus we see that special relativity reveals that time and space are inextricably interdependent. They interact with each other in a very intimate way. In the words of Minkowski: "Henceforth space by itself, and time by itself, are doomed to fade away into mere shadows, and only a kind of union of the two will preserve an independent reality."[4] This is the first important feature of relativity to be stressed in this chapter, and its significance will be compared to both Eastern and Western religious thinking as discussed in Part III.

This interdependence of time and space reveals itself in very specific physical effects. With respect to time in particular,

---

*For those more mathematically inclined the situation described by the transformation equations may be more precisely expressed in terms of what are known as space and time coordinates. The use of the word *coordinates* here is entirely similar to its use in the two-dimensional case of a street map where the vertical and horizontal space coordinates are used to locate a particular street. Thus in relating her observations to Bob's, Andrea's space coordinates in general will depend on both the space and time coordinates in Bob's laboratory, etc.

Andrea will notice that clocks in Bob's frame, moving at a high velocity relative to her, run slower than her own. Bob will notice an identical effect when he compares Andrea's clocks to his. A complementary effect occurs in comparing lengths. Both observers will perceive a yardstick to be shorter in the other's frame than in their own, and by the same amount.

These effects also reveal the interaction of time and space from a somewhat different viewpoint. It turns out that space contracts and time slows, or becomes longer, in just such an amount that the total effect for all four dimensions remains unchanged.[5] However, this interrelatedness of time and space does not imply that they become mixed so that they become in any way indistinguishable. Time is not a form of space. Relativity breaks down the isolation of time and space but not their distinction.[6]

Furthermore, time dilation and space contraction are not considered optical illusions of some kind. The clocks in the frame moving relative to the observer actually do run slower. This has been confirmed experimentally by comparing clocks stationary on earth with ones flown around the world. A small difference in time was detected.

Another very significant time-related effect, implied by the earlier discussion of relatively moving frames, is how some aspects of the notion of simultaneity are challenged by relativity. That is, two events that are seen as simultaneous in one frame of reference can in general be observed as one occurring after the other in another frame moving at an appropriate relative velocity.

As an example illustrating this, suppose Andrea decides to swoop down to earth and "buzz" Bob's laboratory by flying her spaceship at a low altitude directly over it in an easterly direction. Let us assume that just as she is directly overhead two lightning bursts occur, one ten miles west and the other ten miles east of the laboratory. Let us further suppose that Bob observes the light signals from these equidistant lightning flashes to arrive simultaneously. (This of course implies that the lightning bolts must have occurred simultaneously in Bob's frame of reference, since they were equal distances from his laboratory.) However, Andrea would observe the flash from the east (her direction of motion) to arrive before the one from the west; i.e., they are not simultaneous in her frame. This fact and the fact that each can nevertheless measure the velocity of light to be the same is only reconciled by realizing that there are differences in Andrea's and Bob's time and space scales* of just such a magnitude as to make this reconciliation possible. Furthermore the reconciliation in general can be accomplished only if there is a concurrent intermixing of time and space in the right proportions. Thus the idea that simultaneity is relative, which this example illustrates, is closely related to the idea of the intermixing of time and space. It is another feature of relativity that I wish to highlight in this chapter.

The relativity of simultaneity does not mean that we can observe the reversal of cause and effect. No experiment in relativity has been devised that demonstrates such a reversibility. It is the velocity of light that establishes the limits of causal connectibility. There are events presumably happening right now on a star that will not be observed on earth for many years, simply because they transpired at such enormous distances that the finite velocity of light limits how soon news of their occurrence can reach us.

The general situation illustrating the restrictions imposed by the finite velocity of light is shown graphically in Figure 2 which depicts what is usually called the "light cone." The side sectors outside of the cones, marked "Elsewhere," include all the times and positions unreachable by light signals to or from the point 0, our present instant and position in time and space. That is, the distance of a point in "Elsewhere" is too far or the time too short, or both, for the light to reach us at 0. (For a more detailed description of the situation see the caption of Figure 2.) Therefore, the points in "Elsewhere" could only be reached if a signal were to somehow penetrate the thus far unbroken "light barrier" and travel faster than the velocity of light.

The diagram, then, tells us that we can receive light from a far-distant star as long as we are willing to wait long enough. If that signal has not yet reached us, the star is in "Elsewhere"

---

*As specified in the transformation equations.

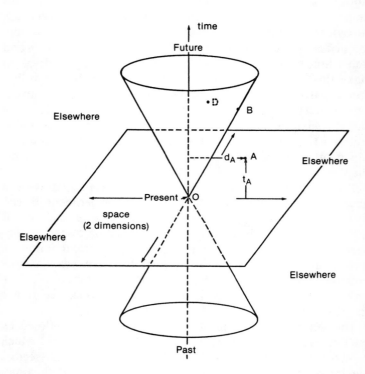

Figure 2. Time is understood to advance upward along the vertical axis, and space (or two dimensions of it) is in the horizontal plane. The limits imposed by the velocity of light are depicted by the surfaces of the cones. The point 0 where the apexes of the cones touch at the center is considered our present instant in time and position in space. The upper cone, indicated by "Future," is all times and positions in the future that can be reached by light signals from point 0. The lower cone, indicated by "Past," is all times and positions in the "past" that could have sent a light signal reaching point 0. From an inspection of the diagram, it is clear that the time t_A at point A in "Elsewhere" is too short and the distance d_A too large for a signal moving at the velocity of light to arrive at point 0. However, at point B at the surface of the cone such a signal would just make it. At point D it would have more than enough time to do so.

with respect to our particular light cone. An astronaut in a spaceship would have her own light cone and could see things somewhat differently if she were moving fast enough relative to the earth.

Although not as apparently time-related as the phenomena discussed thus far, perhaps the most popular relativistic effect is one that is implicit in Einstein's famous and well-known equation $E = mc^2$. This equation says that the energy E and the mass m are equivalent. It is the formula used particularly in subatomic interactions, where often mass actually does entirely convert to energy and vice versa. An example is the simultaneous and total annihilation of an electron and a positron (electron with positive electric charge) into energy in the form of two or three photons (particles of high energy light). The reverse reaction, creation of an electron and positron, also occurs. The relativistic effect implicit in Einstein's formula is that m is in general not the mass as we might see it at rest, but the relativistic mass. That is, the relativistic mass is larger than the rest mass because of the additional energy furnished it by virtue of its velocity. We here on Earth will observe a spaceship to have a smaller mass (in fact its minimum) when it is at rest on the launching pad than when it is in flight to the moon. Thus mass is but a form of energy. As we shall see later in this chapter and in Chapter 2, there is a relation between time and mass or energy.

It is again important to understand that the relativistic effects discussed thus far, e.g., the time dilation, length contraction, and mass increase, are all characterized by their extreme smallness in terms of our everyday experience. This is why they had eluded notice for so many ages. They only become apparent in phenomena involving very high velocities, approaching that of light. For example, even at a velocity of one-tenth that of light (18,600 miles per second), the above three effects involve changes of about one-half of one percent. That is, the time will prolong, the length contract, and the mass increase by that percentage. If we care to go to extremes and consider what would happen as the velocity of light is approached, the time would approach cessation, the length would shrink toward zero, and the mass would approach infinity.

Most relativistic effects, and the attendant high velocities, are generally observed in the world of the "very large" and the "very small," that is, in cosmological phenomena and in phenomena involving atomic and subatomic particles. In our world, "the middle world" of everyday life, most physical events can still be dealt with by using the old Newtonian equations of motion, to which the relativistic equations reduce at the comparatively small velocities we usually experience.

The General Theory of Relativity

Thus far I have limited my description to the special theory of relativity, which gives the relationship of observations made on frames of reference generally moving relative to each other with some constant velocity. Although relative accelerations can be handled by the special theory, a full conceptual understanding of the meaning of acceleration is furnished by the general theory. Thus the general theory, true to its name, not only describes phenomena covered by the special theory, but also deals with relative accelerations in terms of gravitational effects and curved space-time geometry.

Basic to the general relativity theory is the insight that the effects of gravity and acceleration are equivalent, in fact Einstein called this the Principle of Equivalence. For example, if you were sitting in a large, completely sealed box which is resting on the earth, you would not be able to tell whether you were being held to the box floor by gravity or being accelerated upwards by one "g." If the box were freely falling in the earth's gravitational field, you would experience weightlessness because the pull of gravity would be cancelled by the acceleration you are undergoing. Since your acceleration is the same as that of the box, you are "floating" in the box.

More specifically, Einstein was able to correlate the physical properties of gravity with geometrical concepts of curved space first devised by the mathematician G.F.B. Riemann. One can get some notion of what curved space might be like by thinking of the two-dimensional space characteristic of the surface of a sphere, where the shortest distance between two points is no longer a straight line but an arc of a great circle (like the equator) around the sphere.

In his general theory Einstein utilized an analogous four-dimensional curved space-time. In this curved-geometry context, a spaceship orbiting about the earth is thought of as following its curved path, not because the earth's gravity is continually pulling on it, but because it is naturally and freely drifting in a curved space-time geometry. Thus in the words of physicist John Wheeler: "Space tells mass how to move, and mass tells space how to curve."[7] Indeed the whole structure of space-time and its curvature depends on the distribution of matter in the universe.

The general theory gives specific predictions concerning the effects of gravity on certain phenomena. One of the best known is the bending of light which passes in close proximity to a large mass such as the sun. One should remember that photons of light have no rest mass (always moving at the velocity of light, they are never at rest), but they do have mass by virtue of their kinetic energy (energy of motion). Therefore they can be affected by the gravitational pull of a large mass such as a star. Another effect is the very small, but real, precession of the elliptical orbit of Mercury around the sun; that is, the elliptical orbit itself rotates very slowly about the sun. A revolution of this precessional motion is completed every 3 million years.

The principal time-related effect in general relativity is that clocks run slower near larger masses or when subject to larger accelerations. This has been most popularly expressed by the famous twin-paradox thought experiment. If Andrea's twin sister, Carolyn, goes on a long, round-trip space journey, she will return to find Andrea noticeably older. This is because Carolyn has been subjected to accelerations not experienced by Andrea.

The twin-paradox experiment reveals an intimate relationship between time and mass (or acceleration). This observation about the relationship may be further understood by emphasizing what has already been stated about time and mass in relativity at the limit imposed by the velocity of light.[8] That is, time approaches a stopping point and mass approaches infinity as the velocity of light is approached. This is another important feature about time that is revealed by relativity, to be further discussed in Part III.

Some Philosophic Implications

Although the theory of relativity rests today on solid supports of experimental verification, many provocative philosophic implications that can be drawn from the theory are still under discussion. One of the most interesting of these arises from regarding time as the fourth dimension. In what sense and in what degree can time be judged a "spatialized" fourth dimension?

Recall that in relativity theory in order to "spatialize" time so that it could be used in the equations in the same way as the three space dimensions, the time coordinate is multiplied by the velocity of light. Although this does distinguish time from the other dimensions, the resulting mathematical formalism is so beautifully consistent and symmetric and shows such an intimate interrelation of time and space that many thinkers and natural philosophers have tended to regard time as essentially spatialized.

According to this line of thought, if space can be regarded as having extension in three directions, time can be regarded as having extension also. Events past, present, and future are ordered linearly along this time scale. There is a kind of world picture then (seen from any given frame of reference) of each piece of matter, and the events in which it participates, being set in a four-dimensional space-time framework. All of physical history is seen in one fixed sweep. Although writers of this view acknowledge the probabilistic, non-deterministic, features of microscopic events as indicated by the quantum theory[9] (as we will see in Chapter 2), there is a deterministic or static aspect to this world view which time-scholar J. T. Fraser has called a "philosophy of being."[10]

In contrast to the philosophy of being, proponents of what is termed the "philosophy of becoming" see the world from any given local frame of reference as developing in a continual irreversible process of becoming.[11] The present "now" in such a local frame of reference is a marker between the irrecoverable past and the future to become. Here time is looked upon as a dimension much more distinct from the spatial dimensions. In the words of Capek: "the relativistic union of space with time is far more appropriately characterized as a 'dynamization of space' rather than a 'spatialization of time.' "[11] Einstein himself seemed somewhat uninterested as well as ambivalent on the matter, at least in the sense that his statements on the subject seem to vary on different occasions. At one time in the early 1920s he stated: "It is certain that in the four-dimensional continuum all dimensions are not equivalent."[12] However, a few years later he said: "The becoming in the three-dimensional space is somehow converted into being in the world of four dimensions."[13]

In weighing the static and dynamic aspects of relativistic space-time, it may be worthwhile to consider a related observation made by Fraser.[14] He points out that with classical Newtonian physics and the concept of a stationary ether there was, as we have seen, the belief in an absolute space, a fundamental reference frame. However, with relativity theory Fraser feels we must accept the notion of absolute motion where no reference frame is sacrosanct, and where there is a different absolute unchangeable referent, a finite maximum velocity, the velocity of light. It is this finite limiting velocity, measured the same in all reference frames, which imposes on us a fluid, relativized physical existence because, for the most rapid communication between frames, signals must be used which travel at this finite velocity.

Summary

This chapter has emphasized three time-related physical characteristics which relativity has revealed to us and which will be developed further in discussing comparisons with religious concepts. The first is that in relating measurements in two reference frames moving relative to each other, time and space are intimately interconnected, but they remain separately measurable quantities. The second is that the simultaneity of distant events is not absolute, but relative. That is, there is a challenge to the notion of universal, instantaneous simultaneity, but this does not mean that causality can be reversed or reordered within any system of frames of reference intercommunicable with light signals. The third is that mass or acceleration has a direct effect on time, specifically in that clocks on greater masses run slower. Time and mass are interrelated. These relationships of time, space, and mass will find an interesting comparison with such relationships drawn from Buddhist and other religious thought discussed in Parts II and III.

2

Time in the Quantum Theory

If classical scientific thought was shaken to its roots by the theory of relativity, it was equally so shaken by the quantum theory.* By the end of the 19th century, with a few exceptions (then considered insignificant), it was generally believed that Newtonian mechanics, if applied carefully and painstakingly enough, could in principle explain even the most complicated of physical phenomena, such as the motion of all of the molecules in a gas. This deep-seated faith in Newtonian concepts generated as an obvious result a considerable support for the philosophy of determinism. That is, if the motion of every molecule can in principle be known, then everything in the universe including life itself is essentially predetermined. However, after the quantum theory was introduced, if any form of determinism was to survive, it had to be based on a considerably qualified line of logic.

It may be a surprise to some that Einstein did not receive the Nobel prize for his relativity theory but for his particular contributions to the quantum theory. Perhaps it is even more of a surprise that it was Einstein who later turned against the quantum theory and, never able to accept it, tried unsuccessfully for the rest of his life to devise what he considered a more fundamental theory.

His viewpoint was strongly influenced by his admiration for the thought of the great Jewish philosopher Spinoza, who

---

*As was given for relativity, I will give a descriptive survey of the quantum theory, to serve as a basis in understanding the time concepts to be discussed in this as well as later chapters. Again many texts are available for more detailed and comprehensive study at all scholastic levels.[1]

espoused a deterministic view of the world. Einstein maintained his position even after he essentially lost a prolonged and famous debate over the validity of the quantum theory with another of its founding fathers, Niels Bohr. Never willing to abide the probabilistic nature of the quantum theory, Einstein maintained that "God does not play dice." However, despite his efforts the theory still stands on firm experimentally verifiable ground.

The Quantum Theory and the Microscopic World

While the effects of relativity discussed in the last chapter are not noticeable unless high velocities approaching the speed of light are involved, it was found that the effects describable by the quantum theory are only observable for physical phenomena "in the small." That is, there is a departure from Newtonian physics in the study of molecules and all smaller particles (i.e. atoms, nuclei, and subnuclear particles such as mesons, quarks, etc.). In this world of physics, often termed the "microscopic" world, as opposed to the "macroscopic" world of every day practical life, many physical quantities such as energy can assume only certain discrete values and do not vary continuously as we are used to macroscopically. Of course, beyond the macroscopic world of intermediate sizes and dimensions, at the other extreme from the microscopic world is the cosmologic or astronomic world ("world of the very large") of extremely large dimensions about which more will be said in Chapter 4.

To grasp some feeling for the sizes involved in the microscopic world, the diameter of an atom is roughly one hundred millionth of a centimeter; that is, if the atom were the size of a pea, then the centimeter would be the distance between New York City and Pittsburgh. The nucleus in turn is about one hundred thousand times smaller than the atom; so that if the nucleus were now the pea, the diameter of the atom would be the height of the Sears Tower in Chicago plus another fifteen stories.

Some History

It was because of the extremely small sizes characterizing the microscopic world that explanations of such quantum effects as the discreteness of energy values went undiscovered for so long. The first breakthrough came in 1900 when Max Planck was struggling to find a mathematical equation which properly predicted the spectrum of light emitted by what is known as a "black body radiator." (The radiation emerging from the entrance of a coal furnace is essentially "black body" radiation.)

True to the long-standing and accepted theory that light exists only as a wave phenomenon, he tried to find this equation in terms of the wave theory. After much painstaking analysis, he was forced to conclude that light could also exist as quanta or particles of electromagnetic energy, usually called photons. In particular Planck, and later Einstein, showed that the energy of the light was quantized and equal to the frequency of the light times a very small fixed number known as Planck's constant.* Thus, the energy of light cannot have just any value, but comes in minute fixed increments whose magnitude depends on its frequency. Planck's conclusion was so difficult and courageous because virtually all experiments until nearly the end of this 19th century had shown that light could be explained as a continuous wavelike phenomenon.

It was Einstein who gave the quantum theory its next major impetus by exploiting Planck's discovery in explaining the photoelectric effect. Again the explanation came in terms of light, as photons or quanta of energy, impinging on a metal surface and causing the ejection of an electron. It was specifically Einstein's work on the photoelectric effect that was cited in his Nobel Prize award.

At this point for a true perspective it is important to keep in mind that the existence of quantization in nature was not new in 1900. Atoms were known to have discrete masses. The standing waves set up in a child's skipping rope, or in water in a circular pan by a series of evenly-spaced drops falling at the center, can only proceed with discrete frequencies which are integral

---

* This is expressed mathematically by the simple formula $E = h\nu$, where E is the energy, h is Planck's constant, and ν is the frequency. The size of h is about 6.6 x 10^{-34}, where 10^{-34} is a fraction with 1 in the numerator and 1 with 34 zeros behind it in the denominator.

multiples of a specific base frequency. That is, if the lowest or base frequency at which the wave motion can proceed is two oscillations (cycles) a second, then the only other possible frequencies are 4, 6, 8.... cycles per second. Furthermore, the notion that light exists as particles was not new either. It enjoyed credibility, even with Newton (with some reservations), until Young and Fresnel with their studies of light diffraction demonstrated the wave nature of light early in the 19th century.

Waves and Particles

With a century of wave theory dominance, the revelation by Planck and Einstein that light could also behave as a particle produced a major upheaval in the natural sciences. However, on hindsight this perhaps should not have been too much of a surprise, at least in a sense. This is because physicists have been able to observe only two possible modes of energy transport, i.e., via particles or waves.[2] We really have no choice but to describe physical phenomena in terms of macroscopic models using one or the other of these modes, but not both at the same time which would actually be a logical impossibility.

Thus, many physical phenomena exhibit what is popularly termed a wave-particle duality. For example, in understanding sound propagation in a gas, a wave theory is needed, but in dealing with temperature and pressure measurements in the gas, a particle theory is used. Similarly then, electromagnetic radiation or light was found to be characterized by such a wave-particle duality. Depending on what aspect of light was being studied, that is, depending on the measuring device and the conditions under which the measurements are made, either a wavelike or a particle-like behavior is observed, but not both simultaneously. It seems then that light or electromagnetic radiation is inherently more complex than can be understood by using the simple and extreme concepts of wave and particle behavior to which we are restricted in our macroscopic observations.

But the duality did not stop with light. In 1924 Louis de Broglie furnished the key insight that not only photons, which only have mass by virtue of their motion (i.e. zero mass at rest), but also microscopic particles with finite or nonzero rest mass, such as electrons and protons, also exhibit a wave-particle duality. Specifically, he suggested a simple relationship between the momentum (which is velocity times mass) of a particle and its wavelength. This is a companion to the Planck-Einstein equation for energy, and it states that the momentum is equal to Planck's famous constant divided by the wavelength of the particle.*

The wavelike character of microscopic particles was confirmed by Davisson and Germer, who showed that electrons could exhibit a diffraction pattern similar to that for electromagnetic radiation.† The wave-particle duality of such finite mass particles was an indispensable step leading to the development of the mathematical formalism for the quantum theory that was soon to come.

At this point it is natural to ask if there is any unity or coherence to be found with the two pictures, wave and particle. First of all one can at least sense some symmetry in the fact that light or electromagnetic radiation, initially thought to be only wavelike, turned out to be also particle-like, while microscopic matter such as electrons, protons, etc., initially thought to be only particles, also exhibited wavelike properties. It was undoubtedly this sense of symmetry that influenced de Broglie in developing his suggestion. Secondly, one obvious link between the two modes is energy. Whether the instruments for observation are such that they reveal the wavelike or the particle-like character of the radiation, the energy of the radiation must be the same. This quantity has a precise value, as well as physical meaning, in both modes of measurement, and therefore has

---

* This can also be expressed in a simple formula: p or $mv = h/\lambda$, where p is the momentum (mass m times velocity v), λ is the wave length, and h is again Planck's constant.

†Diffraction is a phenomenon in which, for example, light passing through a narrow slit is deflected by the edges producing alternate light and dark fringes on either side of the central light beam through the slit. The fringes are essentially caused by alternate constructive (in phase) and destructive (out of phase) interference of the light waves from the two slit edges.

been useful in mathematically linking the two. As will be seen shortly, a meaningful additional connection is also furnished by the notion of probability wave functions.

Complementarity

Aside from these two connections, the only relation between the two modes of physical behavior is that already implied in the above discussion and first summarized by Niels Bohr in his Principle of Complementarity. That is, the wave and particle modes are mutually exclusive but complementary, and each is necessary in rendering a complete description of a microscopic physical phenomenon. If a given measurement reveals the wave character of the phenomenon, then it is impossible to show the particle character in the same measurement, and vice versa. That is, which one of the two modes is involved when a given measurement is performed is always quite definite, and both can never be used simultaneously. Since the modes are mutually exclusive and one cannot be reduced to the other, for the most complete and coherent picture possible both modes are needed.

As mentioned in the introduction, Bohr's Complementarity Principle has been applied in a generalized form to all kinds of dichotomies, many quite nonphysical and philosophical. For example, there have been many attempts to apply the principle to the apparent dualism between science and religion with the hope of better understanding the relationship between the two fields.[3] The possibility of a more restricted application to the dualism in the religious and physical concepts of time will be discussed in Part III.

Heisenberg's Uncertainty Principle

The Complementarity Principle, as we shall soon see, can actually be deduced from the famous and more quantitatively precise Heisenberg Uncertainty Principle which is at the core of the quantum theory. Briefly stated, it says that the uncertainty in the observation of the position of a particle multiplied by the uncertainty in its momentum can never be less than our friend, Planck's constant.*

The principle tells us that infinite exactness in simultaneous momentum and position measurement is impossible since the product of the two measurement uncertainties can never be reduced to nothing. There is a clear limit then as to how precisely a given pair of such physical measurements can be made, for, although h is a very small quantity, it is not infinitely small or zero. This physical reality has nothing to do with imperfections in the measuring equipment but is an intrinsic fact of nature.

Thus if one wishes to measure the position of the particle with ultimate exactness, this is in principle possible, but only at the expense of simultaneously knowing nothing as to its momentum, and vice versa. If, on the other hand, we are content with partial or imprecise knowledge of the particle's position, then a complementary partial knowledge of its momentum is available to us.

One of the principal reasons for this is that one cannot perform a measurement on an object as small as an atom or an electron without interfering with its motion to some extent. That is, how else can one learn about the motion of such a particle unless, for example, a photon of light is bounced off it so that its behavior can be observed? Thus an observation cannot be made on such a system without affecting it. In other words, bouncing a photon off of an electron is analogous to the relative crudity of trying to bounce a golf ball off a softball in order to determine something about the latter's position or momentum. The golf ball has an effect on the softball because their sizes and masses are comparable.

With some understanding of the Uncertainty Principle, the

---

*Expressed more mathematically, if a particle with momentum p moving in a given direction has a position, denoted by s, as measured along that direction from some established origin, then the principle reads:

$$\Delta s \Delta p \geq h,$$

where Δ is the symbol for the quantitative range of uncertainty in the value of a quantity. So that Δs is the range of uncertainty in the measurement of the position s, and Δp is the range of uncertainty in the measurement of the momentum p. The symbol $\geq$ means "greater than or equal to."

connection with complementarity can now perhaps be clarified. When the position of the object of observation, whether matter or radiation, is accurately known, then since it is so closely localized spatially, it assumes a particle form. On the other hand, if the momentum of the object is precisely known, which means (because of de Broglie's relation mentioned above) that the wavelength is known, then it assumes a wave form. Thus the Complementarity Principle can be seen as evident in the extremes, i.e., position or momentum exactness, of Heisenberg's Uncertainty Principle.

However, perhaps of more direct interest with respect to the subject of time is Heisenberg's companion principle which places similar restrictions on the simultaneous measurement of time and energy.* In essence it says again that the uncertainty in the measurement of the energy of a particle times the uncertainty in the time it exists can never be less than Planck's constant. This energy-time relationship has a direct application, for example, in determining the average lifetime of a nucleus subject to radioactive decay. Suppose the energy of the nucleus is rather precisely known. Then, although it is true that the nucleus can decay at any instant, the period of time over which this instant can occur is of much greater duration than if its energy were less precisely known. This relation between energy and time is the first time-related feature of the quantum theory that should be especially emphasized. Some of its implications will be discussed later in this chapter, as well as in others.

Quantum Mechanics

Acting on de Broglie's insight that matter could also behave as waves, Schroedinger and Heisenberg soon developed separate but equivalent mathematical formalisms of quantum mechanics capable of describing the behavior of microscopic phenomena. Schroedinger's formulation was known as wave mechanics, and Heisenberg's as matrix mechanics.

---

*As with the momentum-position uncertainty relation, this one takes the mathematical form: $\Delta E \Delta t \geq h$ where ΔE represents the uncertainty in the measurement of the energy, and Δt is the similar quantity for the time.

In terms of Schroedinger's wave viewpoint, which is more descriptively amenable, the position of a particle is given by a wave packet (physicists usually call them wave functions), the amplitude of which at any point in space is directly related to the probability that the particle is there.* A typical wave packet for a particle localized in the general vicinity of a point in space can

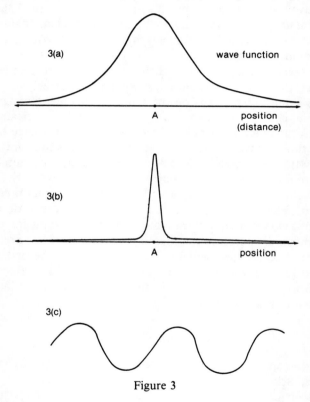

Figure 3

be visualized in one dimension as having the bell shape shown in Figure 3a. Thus, according to the theory, the most probable position of the particle is at the point corresponding to the maximum of the curve (point A in Figure 3a). The farther away from

---

* It is actually the square of the amplitude that is equal to the probability.

that point we look, the less probable it is that the particle will be observed there. If the position of the particle is even more precisely known, then the wave function will be more closely collected or peaked, as in Figure 3b, and its particle nature will be emphasized; if it is less well known, then the function will be more spread out. As it continues to become even less precisely known, the particle begins gradually to develop a wavelike character, as in Figure 3c. As it approaches a clearly defined wave length, (recalling the de Broglie equation which relates wavelength to momentum) the momentum of the particle is becoming more precisely known in accordance with the Heisenberg Uncertainty Principle.

As suggested earlier, a similar relationship pertains between the time or temporal location (i.e. when the particle is passing by) and the energy of the particle. A time wave-packet can give the probability for the temporal location of the particle: the closer the waves in the packet are bunched, the more precisely this location can be determined. If little can be known as to when the particle passes by, then the waves spread out to a more clearly defined frequency, which means, (recalling the Planck-Einstein relation) that the energy of the particle is now more precisely known. This is a further expression of what was cited earlier about the relation between time and energy in the quantum theory.

In our earlier discussion of bridges between particle and wave concepts, it was mentioned that the notion of probability wave function furnished such a link. With the foregoing discussion of probability wave packets, perhaps we may see now how they furnish a smooth link or bridge between the particle and wave concepts of microscopic phenomena. The wave packet can appear as a specifically localized particle or a wave, depending on what measuring apparatus we use for observation, and always in a fashion consistent with the Uncertainty Principle.

Overlapping of Wave Functions

As you may notice in the curve in Figure 3a, the amplitude slowly dies away at the extremes of the wave packet, but the most interesting feature is that it never dies to precisely zero.

This is true of essentially all wave packets or wave functions describing natural microscopic phenomena.

Therefore quantum mechanics tells us that the wave function describing each entity in the universe, at least to some infinitesimal degree, overlaps with that describing any other. The wave function of a macroscopic object such as an apple for all practical purposes dies out within a distance of a few times 10^{-8} centimeters from what is macroscopically taken to be the apple's surface. Such a distance is quite unobservable to the naked eye. Nevertheless, according to the theory there is an extremely minute incremental remainder that stretches out, continually diminishing, to the far reaches of the universe.

One must be somewhat careful here about inferring conclusively from this that we are somehow linked to everyone in the world and everything in the universe. The prevailing body of physical opinion is that such wave functions are only a mathematical representation of a set of statistically related probabilities. In other words, the wave function in Figure 3a could in principle be mapped out by a "thought experiment," in which the position of the particle whose probability the wave function was describing would be measured many, many times. If the number of measurements yielding each given value of the position is plotted against the position itself, then the statistical accumulation of these data would result in a curve gradually assuming the shape given in Figure 3a as more and more measurements are recorded. In gathering these statistics it is clear that any given position measurement will in general not be reproduced on the succeeding measurement, but the accumulated ensemble of measurements yielding the probability pattern in Figure 3a is reproducible.

Despite this very valid probabilistic and statistical characteristic of the quantum theory, it is difficult to escape the thought that there may be some metaphysical meaning to the quantum mechanical prediction that there is a finite probability, however incredibly small, that you, or some part of you, could be found in Tahiti in the winter instead of shivering on a street corner in Montreal. Although this is a fascinating thought, in expressing it we must again realize that this probability is so small that in practical life experience it is effectively zero.

Of course, an argument for such a universal interrelation could also be made through the electromagnetic or gravitational forces. Both of these forces are subject to essentially the same basic inverse square law. For example, there is a very small, but nevertheless finite, force of attraction between you and a rock on Mars, which is directly proportional to the product of your respective masses (yours and the rock's) and inversely proportional to the square of the distance between you and the rock. That is, if Mars were twice as far away, the force would be four times weaker.

These ideas of cosmic interrelations, whether deriving from quantum theory, gravity, or electromagnetism, bear a general conceptual similarity to ideas expressed in Eastern religious thought. In Part III, I will further discuss such ideas, along with those brought up in Chapter 1 concerning the interdependence of time and space in relativity.

Observer Theory

The mathematical formalism based on the quantum theory has been enormously successful in predicting the statistical behavior of microscopic systems. The accumulation of confirmatory experimental results in the study of molecules, atoms, nuclei, and elementary particles over a period of almost sixty years has been extremely impressive. However, though the mathematical framework seems to yield reliable results, some of the conceptual and metaphysical aspects of the quantum theory are still under considerable controversy.[4]

Perhaps the most extensively discussed of such aspects has been the problem of observer theory. As we learned earlier, in contrast to the macroscopic measurements of our daily lives which can be made with miniscule effect on the object measured, a microscopic system cannot escape being affected by an observation. This is because the only device available is something comparable in size to the object being measured. Thus the observer is no longer objectively detached from the observed as seems to be the case in the macroscopic world: observation is now an interactive process. Where do we draw the line between observer and observed? It is because of this interaction between the two that John Wheeler in his provocative discourses on the subject substitutes the word "participator" for "observer."[5] But regardless of such labels, when is the interactive process of observation or participation complete?

Of the early formulators of the quantum theory, it was Niels Bohr who probably gave the most extensive thought to those questions. According to Bohr, no "elementary phenomenon is a phenomenon until it is a registered [observed] phenomenon." Wheeler's version of this is perhaps more specific: "No phenomenon is a phenomenon until brought to a close by an irrevocable act of amplification."[6] The act of amplification is of course the act of our measurement with some macroscopic instrument, which is used to amplify the interaction with the observed microscopic event so that we can register or record it. On the other hand, Bohr went so far as to claim that no phenomenon was valid until it had not only been recorded but also reported to someone else. We can therefore see that where to draw the line between observer and observed, or even whether an attempt should be made to make such a demarcation, might be a matter for considerable discussion.

It may be apparent that the observer problem is related to the epistemological problem discussed by philosophers as to the reality of events occurring independent, or beyond the range, of our consciousness. This, of course, suggests the obvious possibility of circumventing the problem by setting up automatic detection and recording instruments to make observations while we are off doing something else. But taking this recourse into account, we can ask whether or not a phenomenon that we have neither observed directly with consciousness or indirectly with automatic devices is really a phenomenon. According to the quantum theory it is not, because all the theory is really telling us is the simple fact that we cannot say anything physically about a phenomenon until we have performed a completed measurement on it.

Despite these questions as to the nature of the relation between observer and observed, it still remains true that measurements that are *statistically* reproducible can be made on microscopic phenomena with macroscopic instruments. But again, as Heisenberg warns us, in so doing we are not observing nature itself but nature exposed to our method of questioning, i.e., to the instruments available to us.[7]

This method of questioning, this irrevocable act of amplification, can *in a sense,* give us the power to influence the outcome of microscopic events that were initiated long ago. Let us see how this may be so. Before we measure, say, the position of a particle, all we know is that the probability that the measurement will yield a certain result is given by some wave function or pattern, an example of which was given in Figure 3a. But once we have made the measurement on the phenomenon and have a fixed, specific, numerical result,* we have by the measurement process influenced the outcome of the phenomenon, due to the interaction of our instruments with it.[8]

Wheeler pursues this line of thought in discussing some of the implications of making measurements on the primordial photons generated in the Big Bang, of which more will be said in Chapter 4. Wheeler points out that in making such a measurement, or any microscopic measurement, "the answer we get depends on the question we put, the experiment we arrange, the registering device we choose.... We are inescapably involved in bringing about that which appears to be happening." Our choice of device "has an irretrievable consequence for what we have the right to say about the past, even the past billions of years ago, before there was any life on earth or anywhere else." Thus in a strange and indirect sense, but only in a sense, we in the present may be in some degree "making the history" of something that happened billions of years ago. So that to Wheeler: "This circumstance destroys the view that the past exists 'back there.' The past has no existence except as it is contained in the records, near and far, of the present, and the same applies to the presents to come, the future."[9] Somehow we are interacting with time.

Space and Time Quantization

Another aspect of time and the quantum theory which has been of interest to some physicists is the possibility of space-time

---

*Again remember, it would take many such measurements to develop a statistical probability pattern such as shown in Figure 3a.

quantization. Are space and time on a very submicroscopic level divided into infinitesimal quantized increments? Since other physical quantities such as electric charge and energy are quantized in the microscopic world, why not space and time? The idea is still subject to study, and although the theory is presently receiving mixed support, it must be said that it has not yet (as of this writing) been conclusively proven or disproven.

Indeed it begins to be difficult to prove or disprove any theory, especially by experiment, when we attempt a look at the minimicroscopic world beyond the limits imposed by the Heisenberg Uncertainty Principle. For example, theoretical physicists are now working with theories that hypothesize what might be happening within such incredibly short distances as 10^{-33} cm. Recall that Heisenberg's principle states that the uncertainty in the position of a particle times the uncertainty in its momentum can never be less than Planck's constant. So that to examine the spatial characteristics of phenomena with an uncertainty less than 10^{-33} cm would mean dealing with an enormous complementary uncertainty in the momentum. It would require an accelerator as large as our galaxy to accelerate particles with a momentum high enough to overcome this uncertainty and probe such short distances.[10] Congress would never pass the appropriation bill for such an accelerator!

In any event since this is impossible, theoreticians are devising theories which might describe the behavior of space-time in this minimicroscopic world. They are using as one of their guides experience gained in developing theories that have been consistent with the experimental results from the high-energy accelerators which do exist. In this infinitesimal realm, according to theorists, the vacuum (which may appear to be nothing at the macroscopic level) contains some minimum concentration of energy that occasionally can accumulate to create unobservably short-lived particles (often called virtual particles which vanish before detection). The energies (or masses using Einstein's $E = mc^2$) of the particles and their lifetimes are always such that the Uncertainty Principle is obeyed. This behavior is generally called quantum or vacuum fluctuation. Wheeler's name for this turbulent nature of the vacuum at the theoretical submicroscopic level is the "quantum foam." He speculates that the curvature of space-time (discussed in Chapter 1) in this realm can assume a variety of weird and distorted shapes. He also points out that at the incredibly short distance of about 10^{-33} centimeters, time in fact may have no meaning.[11] There maybe a total breakdown of all concepts of "before and after."

The foregoing theoretical conjecture visualized by Wheeler is not the only situation in which we will find time may have no meaning. As we will see in Chapter 4, it is not at all clear that previous to the Big Bang there was any such thing as time. It is again Wheeler who uses these arguments to point out that in an ultimate sense the whole concept of time, at least from his point of view, is in trouble. He wonders if some deeper theoretical framework is needed, a framework that transcends time.

More on Wheeler's thought will be treated in Chapter 4. It may be clear from what has already been presented that there is a virulent and restless probing for new approaches in the theoretical physics community today, some of which are related to space-time quantization. An interesting and relevant example of this is the recent theoretical work of Nobel Laureate T. D. Lee, who is questioning the time-honored belief that continuous mathematical functions can accurately represent physical phenonomena.[12] All of calculus is based on the assumption of the continuity of such functions, the simplest case of which would be a straight line in which all of the points are infinitesimally close with no breaks in the line, however small. Lee maintains that such a concept does not correspond to physical reality and should not be used to describe that reality.

Consequently, on the assumption that only a finite number of discrete observations can be made in a volume of physical space, he is undertaking the program of reformulating, among other things, all of the equations of classical and quantum mechanics in terms of what he calls a discrete (or discontinuous) mathematics. This entails the visualization of a kind of latticelike four dimensional space-time. (For those who know some calculus, this means using summations of discrete elements instead of integrals. Lee holds that it is the discrete mathematics (using summations) instead of continuous mathematics (using integrals) that truly represents physical phenomena.) For him continuous mathematics only represents an approximation, which is just the opposite of what has been believed in the physical science community for the last 300 years.

Naturally his discrete mathematics must at least match the traditional continuous mathematics in its ability to predict physical phenomena. The hope is that it might do more than that and help us understand some phenomena that the latter mathematics fails to describe accurately.

Determinism

The intrinsic indeterminacy that characterizes the microscopic world, as we mentioned at the beginning of this chapter, destroyed all notions of the idealized mechanistic determinism that had such support at the end of the 19th century. The extreme of this idealization essentially entailed the belief that if the position and velocity of every particle in the universe could be known at one instant, then at least in principle, the future of the entire universe could be predicted. However, quantum mechanics now tells us that if any predictions are to be made, particularly at the microscopic level, they are at best statistical averages. Even without quantum mechanics, such an extreme mechanistic determinism is virtually insupportable. As Feynman points out, under such a theory the slightest error, in say, the position of one atom in a gas, would by successive collisions be magnified enormously.[13] Thus even classical mechanics, from any kind of practical point of view, is indeterministic.

Summary

Let us now summarize the time-related subjects treated in this chapter, many of which will be discussed later in finding links to religious views of time. We saw that there is a complementary connection between time and energy in the microscopic world that is evident in the operation of the Uncertainty Principle.* It

---

*It is interesting to remember in the previous chapter that in the cosmic world, where the effects of gravity are especially apparent, a connection between time and mass (which through $E = mc^2$ is equivalent to energy) was suggested. These connections between time and mass-energy will be discussed further in Part III.

was noted that from the position and momentum relation of the Uncertainty Principle the Complementarity Principle, dealing with the wave-particle duality, could be deduced.

The Uncertainty Principle is based on the fact that no microscopic event can be measured without affecting it by the interaction with some other microscopic mechanism used as the measurer. Thus the observer is no longer objectively detached, as he can be for all practical purposes in the macroscopic world where with a sensitive enough device the object of measurement is practically unaffected. This interaction between the observer and the observed led Wheeler to use the label "participator," and to point out how, as such, we have an irrevocable influence on the result of a microscopic measurement by virtue of the particular instruments and settings that are utilized. A striking example of this is how *in a sense* we influence the past of billions of years ago by present measurements made on the primordial photons from the Big Bang.

The remarks on the possibility of time and space quantization serve as a starting point for the discussion of quantum fluctuations and submicroscopic effects occurring within dimensional domains much smaller than are practically measurable according to the Heisenberg Uncertainty Principle. It is in such incredibly small regions where Wheeler points out that time in an ultimate sense may be a meaningless concept.

Finally, two subjects indirectly related to time were treated: wave function overlap and determinism. In the former case we saw that there is an infinitesimal but nonzero probability for everything in the universe to be everywhere in the universe. In the latter, it was shown how the Uncertainty Principle and the probabilistic features of the quantum theory make any notion of an ideal classical determinism impossible, as it is anyway from a practical point of view, using the arguments of Feynman.

3

Reversible and Irreversible Time

It is generally taken for granted in our modern life that time flows in one direction and does not reverse or flow backwards. Spilt milk does not leap up from the floor and collect back in the glass. A skier does not reverse direction and slide up the mountain. Given such facts of daily experience in our macroscopic world, it may seem at least a waste of time if not ridiculous to be thinking about time-reversal. Indeed this would be true were it not for the fact that in the microscopic world, and even in some valid descriptions of macroscopic phenomena, the notion of time reversibility turns out to be a very useful and meaningful concept.

Reversible Time

Some sense as to the validity of such a concept may be gained by considering a collision of two balls on a pool table shown schematically in Figure 4. A collision with the arrows reversed is a perfectly possible collision among the many that occur in the game, and this situation is described by the same physical law of motion as the original collision. That is, if each ball came in from its exit direction instead of its original entrance direction, a perfectly acceptable collision would occur, i.e., the reverse of the original.

Perhaps the most convincing way to grasp the notion of reversibility in the above example would be to take a motion picture of the collision between the balls. Suppose the film were taken off the spool and thrown on the floor in a tangled mess.

How, in principle, would one know which way to wind the film on the spool again. Either way a perfectly valid, mathematically describable physical collision would be shown by the projector, one collision being the reverse of the other. (This part of this example is due to D. Park.[1])

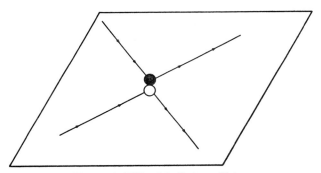

Figure 4: Billiard balls in collision.

Although there are some small frictional heat losses of the energy of the balls to the felt on the pool table, the example at least approximates an ideal frictionless collision. Even more ideally time-reversible collisions are experienced, for example, by molecules in a gas. Here particularly the motions involved in individual molecular collisions can also be described in a way that is completely symmetric in time.

This time symmetry is reflected in the mathematical equations the physicist utilizes to represent the motion. That is, regardless of whether the physicist uses $+t$ (denoting forward time) or $-t$ (denoting reverse time) in the equations, the form of the equations does not change. A perfectly valid physical solution is obtainable either way.

Indeed, excluding one exception to be discussed later, this is true of all interactions, between all particles in the microscopic world. This is an extremely powerful statement and worth thinking about for a while. Any of the electrons, protons, nuclei, atoms, and molecules that make up us and the rest of the universe, when considered in an individual collision or interaction with some other microscopic particle, can be described in this time-symmetric fashion.

A particularly interesting example is exhibited in the study of the interaction of electrons with positrons. It turns out that the theory quite adequately explains these interactions under the assumption that the positron behaves as if it were an electron traveling backwards in time. There is a balanced and reciprocal behavior of the two particles. As far as the mathematics is concerned, the positron with its positive electric charge proceeds in the negative time direction, while the electron with its negative electric charge moves in the positive time direction. The electron-positron interaction is then a prototype for the beautiful symmetry with respect to time that is revealed in the microscopic world.

Perhaps it is now understandable why so much thought has been given to the unidirectionality of time in our daily macroscopic world. For the profound question immediately arises: How is it that the phenomena in the macroscopic world are characterized by irreversible time, while the individual microscopic events comprising this macroscopic world are amenable to a time-reversible or time-symmetric description? A little thought should indicate that this is far from a trivial problem. The answer is not at all obvious even intuitively, and many aspects of it are still subject to controversy today.

From Microscopic Reversibility to Macroscopic Irreversibility

It was the statistical theory of thermodynamics, developed in the last century largely by Ludwig Boltzmann, that yielded the first generally accepted answer to the question. He was able, from the statistical analysis of the motion of molecules in gas, to derive mathematical expressions for such thermodynamic quantities as temperature, something which is commonly measured for all kinds of macroscopic objects, from humans to electric power transformers. He showed that the macroscopic quantity known as the temperature of an object was directly related to a statistical averaging of the microscopic motion of the molecules of the object.

However, Boltzmann also derived the statistical counterpart for another thermodynamic quantity known as entropy, which is relevant to the time "reversibility-irreversibility" question. In essence he related the thermodynamic entropy of a given isolated object or system of objects to the degree of molecular disorder in the system. As the disorder or loss of organization of a system increases, so does its entropy, the measure of its disorder.* It is this progressive entropy increase that many thinkers associate with irreversible time.

With some exceptions which will be treated later, a vast body of physical processes in the universe exhibit a steady advance toward disorder or entropy increase. Heat flows from a hot body to a cold one, never to be recovered unless by some kind of machine (e.g., a heat pump) that will itself expend more energy in the effort to recover the energy lost to the cold body than the amount it retrieves. A drop of ink placed in a glass of water will soon uniformly color the whole glass and is not expected to organize back into a drop again. The probability is extremely slight that a pack of cards originally ordered numerically and according to suits, once shuffled, will return to the initial order in any tolerable length of time. The refined and ordered configuration of the molecules of gasoline is disintegrated in the burning of the fuel in the engine cylinders of a car. The energy produced is not all transmitted to the order implicit in giving the car motion and direction. Much is irrevocably lost in the exhaust and heat radiated from the engine block.

It is obvious that there are endless examples of processes with a net gain in disorder or entropy and concurrent loss of order and organization. However, the most important characteristic of such entropy increase is its irreversibility, which brings us back to Boltzmann. He was able to show, despite the time-reversible nature of individual microscopic interactions, that the cumulative statistical randomness of these individual events add up to irreversible behavior of the gas at the macroscopic level. When it is realized that in a liter of air that we breathe there are of the order of 10^{23} molecules (1 with 23 zeros behind it!),

---

*Denbigh makes an interesting distinction between order and organization. For example, a crystal lattice possesses order while a living cell is organized. In the general treatment given here this distinction is not made. (K.G. Denbigh, *Three Concepts of Time* [New York: Springer-Verlag, 1981], pp. 105 ff.)

perhaps such resultant statistical chaos is understandable. With this in mind, consider one last example. If a small container of one gas is placed in the corner of a larger container of another gas and the smaller vessel is opened, the two gases will soon homogeneously mix due to billions of time-reversible collisions; the first gas is very unlikely to collect again in the small container. Thus time-*irreversible* macroscopic behavior is derived from time-*reversible* microscopic behavior.

Much of what has been said here is summarized in, and can be inferred from, what is known as the Second Law of Thermodynamics. In essence this law states that in any isolated physical process involving energy exchange, there is always some energy irreversibly lost to the surroundings; i.e., there is always an increase in the entropy of the system. This law is consistent with the First Law of Thermodynamics, which states that in all interactions energy is conserved. This is because even though energy may be lost to the surroundings, it is not lost in a total sense; if one considers the larger system which includes the surroundings, the First Law still holds. The "lost" energy that has gone to ultimately heat up the surroundings is not actually lost but is irretrievable for any further practical use.

It is interesting to note that Erwin Schroedinger, mentioned in the last chapter as one of the founders of the quantum theory, has found it especially significant that irreversible time flow is not worked into the physical mechanism of microscopic events but is essentially a macroscopic statistical effect.[2] He feels that this realization, along with the intermixing of time and space which is characteristic of relativity theory (discussed in Chapter 1), constitutes a relaxing or softening of the rigid Newtonian concept of relentless, independent time flow. To Schroedinger this constituted at least some kind of incremental consolation that we were not entirely subject to the tyranny of "Time."

However, we need to investigate in more detail the matter of time irreversibility in the framework thus far outlined. Recall that Boltzmann found it practical, if not necessary, to use statistical techniques in passing from consideration of individual molecular events in a gas to the thermodynamic behavior of the gas as a whole, comprised of astronomical numbers of molecules. However, to some readers there still may remain the haunting speculation that, at least in principle, if there were a computer large enough, we could keep track of the trajectories of all of the trillions and trillions of gas molecules and thus achieve a very precise prediction of behavior of the collective assembly in the container, certainly more precise than the statistical averaging techniques mentioned.

Such a speculation brings us back to the problem of determinism discussed in the last chapter. There we saw that despite the classical idealized notion that in principle we could predict all of the trajectories in the gas, the Heisenberg Uncertainty Principle told us that both the position and momentum (or velocity) of a microscopic particle cannot be simultaneously measured with arbitrary accuracy. This being the case, it is perhaps understandable that statistical methods must be used in describing the gross macroscopic behavior of the gas.

But let us set aside the Heisenberg Principle for a moment and for amusement press the statistical arguments to their limits. Even if we grant that it is extremely unlikely that the playing cards when reshuffled will return to their original order, or the drop of ink to its original concentrated volume separated from the water, or the molecules of a gas to their original configuration of positions and velocities, there is nevertheless a calculable probability, however extremely infinitesimal, that such repetitions will occur.* The length of time it would take for such a reoccurrence to transpire, of course, depends on the complexity of the system under consideration, but it can be said that for systems of macroscopic size, the time would amount roughly to an incredible number of units, each of some 18 billion years, the present life of the universe.

The reason for bringing up the matter of what might be termed "recurrence of physical states" is that if such is the case, then how can the Second Law of Thermodynamics be valid? If a physical system can return to its original state, then its entropy (disorder) has not increased continually but has had to decrease to accomplish such a return. Accordingly it would also be difficult to associate unidirectional time flow with the progressive entropy increase characteristic of the Second Law.

How does such an association of irreversible time with entro-

---

*This interesting point is discussed more fully by Denbigh. (*Three Concepts of Time.*)

py increase have any validity if a system can ultimately return to its original state? First, the time it would take for such a return is so incredibly long that for all practical purposes, it is never. Second, how do we know, coming back to consideration of the Heisenberg Uncertainty Principle, that all of the molecules of the system would be in exactly the same state they were initially?

Therefore, although many subtle aspects of the matter are still under very active discussion, there is considerable credible support for the association of the irreversibility of entropy change with the irreversibility of time. However, as Schlegel points out,[3] in making such an association we must be careful to not fool ourselves into the belief that we have now discovered what Eddington terms the "arrow of time,"[4] or even its cause. For men and women were aware by intuition, some of it religious, of the unidirectional flow of time long before they ever heard of entropy.

Other Arrows of Time

Now that we have brought up the subject of humans: what about them and the Second Law? Do they not constitute a violation of it? A human is certainly an incredibly ordered system (depending perhaps on your sociological viewpoint). The growth of an infant into a logical, thinking adult involves a lot of entropy decrease or accumulation of order; and, of course, the same is true of any animal or plant.

The answer to this apparent paradox is that humans do indeed constitute an apparent local violation of the Second Law. But this is only apparent because in the life process living organisms cannot be isolated from their surroundings, from which they obtain food. Thus the proper application of the Second Law must include the person's surroundings as part of the system to which the law is applied. If, for example, one were to include all of the food that had been "disordered" into the excreta of the person, the chances are that there would be a net gain in entropy[5] (although to my knowledge experiments to prove this have not been performed).

Despite the fact that, if the total isolated system is chosen correctly, the Second Law is not violated, the existence and growth of such apparent local violations as humans, animals, and biological organisms are taken quite seriously by some thinkers as another process in nature with which to identify unidirectional time. These copious examples of entropy decrease (order increase) or, as Layzer puts it, information increase,[6] constitute another representation of irreversible time which he calls "historical time." Thus for Layzer there exist concurrent "arrows of time": the "thermodynamic arrow" characterized by the general tendency for many systems to be subject to progressive disorder (entropy increase), and the "historical arrow" characterized by growth of order or information in a system.

Now suppose a system continually increases its entropy until it finally reaches thermal equilibrium or maximum entropy. Then, for those who believe that it is entropy increase that constitutes unidirectional time itself, time for that system would stop. This interesting observation has often been applied to the universe as a whole. Some are of the opinion that the universe on the average is tending toward thermal equilibrium, which it may approach billions and billions of years from now. Of course, this is a vast averaging, because there are many, many counter examples for the growth of order, e.g., ourselves and animals already cited, as well as beautifully ordered silicon or salt crystals, not to mention spinning galaxies and solar systems. On the other hand it is the thesis of Layzer that the universe started at thermal equilibrium and is expanding in a way that accommodates both entropy and information growth in local subsystems.

Another cosmological question that some theorists conjecture may have an influence on time is whether the universe is open, continuing to expand forever, or closed so that at some future date it will reverse and start contracting. Although this question will be discussed more completely in the next chapter, it is relevant to note here that Andrei Sakharov, famous Russian dissident and Nobel laureate, has postulated a model of the universe in which the arrow of time reverses when cosmological expansion reaches its maximum and reverses.[7] He actually theorizes a cyclical universe in which the expansions and contractions alternate, and presumably along with them, the arrow of time (which associates time with the evolution of the universe).

The status of such cosmological theories, and thus the nature of what has been coined the "cosmological arrow of time" is constantly subject to change and amendment as more astronomical data as well as data from subnuclear processes are accumulated. In recent years it has become increasingly apparent that time in the cosmologic world is strongly affected by subnuclear phenomena in the microscopic realm. Such phenomena, according to elementary particle theorists, played a vital role in the very early evolution of the universe, e.g., the first three minutes after the Big Bang.[8] Again, more on this subject will be described in the next chapter, but it is important in this discussion of time's arrow to discuss a particular elementary particle phenomenon. This is because it is the exception, referred to early in this chapter, to the statement that all phenomena in the microscopic world can be described by a time-reversible theory.

This one exception is found in the decay process of a very short-lived particle known as the neutral (no electric charge) K-meson. This is one among many mesons which are particles of mass intermediate between light particles such as electrons and heavy particles known as baryons, the lightest of which are the proton and neutron. Although the theory of neutral K-meson decay is beyond the scope of this book, basically what the experiments revealed was that the decay proceeded by two modes. One mode yielded three particles: a negative pion (a lighter meson), a positron (a positively charged electron), and a neutrino (a particle of zero or near-zero mass); the other yielded the antiparticles of the above three: a positive pion, an electron, and an antineutrino.

It turns out that if this decay process were describable in a completely time-reversible way, it would proceed with equal probability by the two above decay modes. It almost does, but not quite. There is a small but measurable violation of time symmetry of about one part in one hundred thousand. At first sight this very small exception may seem insignificant. In many respects it might be except for the work in recent years by elementary particle theorists in attempting to describe the phenomena occurring shortly after the Big Bang. In such a regime of ultra-high energies, the departure from time symmetry was probably larger than it is today. However, even a relatively small violation of time reversibility, according to theorists, could set up a chain of events that would lead from a universe consisting of equal numbers of particles and antiparticles to one having mostly particles, which seems to be the case today.[9]

If the theory holds up on this line of reasoning, it seems to me that this result is a matter of fundamental importance in establishing not only a "universal or cosmologic" arrow of time, but also "thermodynamic" and "historical" arrows. For if it were not for the imbalance that made possible the preponderance of matter over antimatter, there would be no protons, neutrons, and electrons to make up, for example, the molecules in a gas that exhibits the thermodynamic arrow of time. Nor would there be evolving biological orders of molecules to exhibit the historical arrow of time, not to mention the fact that we would not even be around to think about it.

Is There a Unity of the Time-Reversible and Irreversible Realms?

This important exception in the generally time-reversible microscopic domain, along with time irreversibility of the macroscopic domain, prompts one to wonder whether there is a means of looking at nature, a theory which somehow includes both reversibility and irreversibility in a way that bridges the microscopic and macroscopic worlds. As mentioned early in this chapter, a clear-cut construction of such a bridge from time reversibility to irreversibility is not a trivial matter. Indeed according to Prigogine, who among many others has gone beyond the pioneering work of Boltzmann and given extensive study to this subject, "it is perhaps one of the hottest problems of our time, one in which science and philosophy merge."[10]

Prigogine, a Nobel laureate, sees the physical world describable by classical and quantum mechanics as one which has been made idealized and static by these theories, theories strongly influenced by Greek and Newtonian concepts. To him such descriptions are static because the direction of time essentially makes no difference, that is, it is reversible and symmetric. He further claims that these static or time-reversible theories have dominated our perception of how to describe physical nature so that time-irreversible theories have often been regarded as mere ad hoc appendages. They have been added to the traditional theoretical framework as phenomenological or little more than sophisticated engineering equations with little deep conceptual content.[11]

Prigogine thus champions irreversible thermodynamic processes as playing an equally fundamental role in the physical world and maintains that irreversibility commences where classical and quantum theory reach their limits. He feels that the essentially static descriptions afforded by classical and quantum dynamics represent a world of "being," while irreversible thermodynamics, because of its inherent directionality, reveals a world of "becoming." The principal mission of his book *From Being to Becoming* is to show that irreversibility is not described by some supplementary assemblage of mathematical relations, but that both reversible and irreversible phenomena are embraced in a much larger mathematical formalism, which is a generalization of the traditional "static" formalism.[12]

Thus he develops equations in his theory that contain both a reversible and irreversible component.* In his theory time becomes a much more clear-cut observable quantity than in conventional quantum mechanics. Indeed Prigogine suggests that philosophically his theory promotes time from being a mere mathematical label in classical and quantum dynamics to being endowed with a new meaning associated with evolution.[13]

One of the more interesting insights that Prigogine shares in the development of his theory is that any measurement made on a physical body involves some irreversibility. More specifically, the entropy of the entire system of measured body and measuring device (probably also including the observer) has been increased in this process.[14] This is consistent with the point made in the last chapter that no measurement on a system, especially a microscopic system, can be made without disturbing the system. Thus it is sensible that at least some incremental irrecoverable energy loss might result in the measurement process. In any event the implications of Prigogine's insight on observation

---

*This is actually true of Boltzmann's early work in which his so-called H-theorem possessed this property. Prigogine's equations are intended to be more general.

theory in quantum mechanics (which theory was discussed in the last chapter) have yet to be fully explored. However, further speculations related to these implications will be presented in Part III.

Summary

This chapter was introduced by discussing the concept of time reversibility which, despite our notions of the irreversibility of time in daily macroscopic life, characterize the descriptions of almost all microscopic phenomena, as well as some macroscopic phenomena that can be handled with simple classical mathematics. This was an important preamble to the treatment of entropy and disorder increase because one of the central problems we encountered was how an assembly of individual time-reversible events could statistically result in time-irreversible phenomena at the macroscopic level. Two important outgrowths of this discussion were that the concepts of physical determinism and the statistical possibility of a system returning to a previously ordered state after an incredible time, while both ideally logical, were not considered valid. Among reasons given, perhaps the most telling was the argument that all systems are comprised of microscopic particles and thus adhere to Heisenberg's Uncertainty Principle. Therefore, any notion of determinism would have to be modified to some probabilistic form and the possibility of exact return would always be uncertain.

The idea of associating unidirectional, irreversible time with entropy (disorder) increase was introduced as the thermodynamic arrow. The historical arrow, associated with entropy decrease (increase in order as in crystals, plants, animals, humans, etc.) was then discussed, as well as the cosmologic arrow, associating irreversible time with evolution of the universe. In the case of thermodynamic time in which entropy increase dominates, it was suggested that if a system reaches maximum entropy (thermal equilibrium) then there would be no identifiable changes of physical state with which to associate a flow of time. This notion was extended to the speculation that the expanding universe might ultimately cool and approach such a state.

The one exception to time-reversible phenomena in the microscopic world, the decay of the K-meson, was then described and the fundamental importance of such small departures from reversibility in the early universe was emphasized as crucial to the very existence of unidirectional time. This was followed by a presentation of the thought of Prigogine and his attempts to develop a unified theory encompassing both time-reversible and irreversible processes. Many of the subjects I have dealt with in this chapter as well as the two earlier chapters will play a role in the next and final chapter on physical concepts of time.

4

Focus on the Universe:
Cosmology and Elementary Particle Physics

In recent years there has been a progressive trend toward multidisciplinary approaches to many problems, whether scientific, economic, or political. Indeed this book represents such an approach with respect to time. Even within a discipline such as physics, the knowledge and expertise of two or more subfields is being brought to bear on problems with increasing frequency. This has been the case particularly in the last ten years with cosmology and elementary particle physics,* previously considered quite disparate subfields of physics, now jointly focusing on the problem of a detailed description of the very early history of the universe. The philosophic and/or religious import of the physics "in the largest" (cosmology) and "in the smallest" (elementary particles) coming together to understand creation has yet to be fully assessed.

These two lines of endeavor have converged on developing and testing important details of some of the most credible current theories describing the gross behavior of the universe, the most popular of which is the Big Bang Theory (BBT).

---

*This area of physics deals with subnuclear phenomena characteristic of the increasing array of particles still being discovered, mostly short lived, which range from the massless or nearly massless neutrino to masses considerably larger than the proton and neutron. This field studies the particles and their interactions with each other. It has provided us among other things with the quark theory, which postulates that, in addition to such particles as electrons and neutrinos, the building blocks of a large portion of matter in our universe consist of relatively few kinds of quarks and gluons (the latter to "glue" the quarks together).

Perhaps the single most important feature of this theory is that there was an instant at which the universe began; it, as well as time itself, may not have been around forever. This fact is in some respects consistent with many creation myths (in particular the myth upon which Genesis in the Old Testament was based) and the insights of St. Augustine, and it will be discussed in Part II.

Some Basic Features of the Universe

Based on the experimental observations of astronomers and physicists in the last sixty-five years, there are several fundamental characteristics of the universe which need to be explained by any successful theory, in particular the BBT. Probably the most interesting of these was observed by such astronomers as Slipher, Hubble, and Humason, who showed that distant galaxies are receding from us at enormous speeds. They were led to this conclusion by measurements which revealed that the wavelength of the light from these galaxies was shifted by the Doppler effect to longer wavelengths, i.e., toward the red, or lower frequency, end of the visible light spectrum. The Doppler effect is evident, for example in the sensation of higher and lower pitched sound of an automobile horn as the vehicle approaches and then recedes from the listener. It is the same with the light from the receding galaxies;* the higher the recession velocity the greater the red shift, or shift to lower frequencies and longer wave lengths.

The Doppler effect made it also possible for astronomers to note that the farther away a galaxy or star was, the faster it was receding. Stars at the edge of the observable universe, as of this writing, have been recorded to be moving at the incredible speed of about 93% of the velocity of light! If one devotes a moment to visualize this situation, it has the general properties of an

---

*There is another possible cause for the observed red shift. From Einstein's general theory of relativity it is known that light from a source in an intense gravitational field, for example one furnished by a very massive star, will be shifted towards the red. However, the prevailing opinion is that most red shift observations result from the Doppler effect.

explosion with the farthest fragments moving the fastest. Such obvious comparisons with a classic explosion undoubtedly led George Gamow and others to propose the BBT. The theory tells us, as suspected, that the universe came forth in a burst of enormous energy and is expanding at a prodigious rate.

However, in making a comparison with an explosion it is easy to fall into the error of visualizing ourselves as being at the center of it. A much more accurate way of visualizing the expansion resulting from the Big Bang is to compare it with the two-dimensional case of the surface of a balloon dotted with ink spots. As the balloon is blown up, all of the dots recede from each other; and the farther apart any two dots are, the faster they are receding from each other. In a two-dimensional universe we on Earth would be just one of the dots, a very small one at that.

In any case it seems apparent that in the beginning the universe was far more compressed than now, and space was also. For in Chapter 1 it was explained that the gravitational effects described by the general relativity theory are such that matter causes space to curve. Recall Wheeler's words: "Matter tells space how to curve, space tells matter how to move."[1] The Big Bang occurred everywhere in this very compressed initial state of the universe, for there was no space outside under such conditions because all space was curved and confined by the concentrated matter therein. Thus in the explosion all parts of the universe began separating from one another like the ink dots on the two-dimensional balloon surface; concurrently there was expansion of space. Because of this continual progressive expansion, the universe in a very real sense is itself a clock, which helps define what was termed the "cosmologic arrow of time" as discussed in Chapter 3.[2]

Since its proposal in 1946, the BBT has received additional support. For example, the theory predicts that about 75% of the universe consists of hydrogen, 24% of helium, and 1% of all the rest of the elements. Experimental evidence confirms this. But the most dramatic experimental support came in 1965 with the discovery by Penzias and Wilson of very low-energy radiation (with wavelength comparable to radio waves) coming from all directions. This remnant background radiation is the afterglow of the primordial creation explosion and as such is distributed uniformly throughout the universe. Because of the finiteness of the velocity of light, radiation reaches us now from distant parts of the universe, radiation that was emitted at a very early stage in evolution. Thus we are in effect looking back in time to one of the early epochs following the creation event. At that time the radiation was about 1000 times more energetic, but it has lost this energy in exerting outward explosion pressure, which furthered the universe's expansion.

The existence of this radiation in itself was a prediction of the BBT, but even more convincing was the fact that the spectrum of frequencies was just that expected from the Big Bang explosion. That is, it was essentially the spectrum of frequencies characteristic of a black body radiator such as studied by Planck and discussed in Chapter 2.* Thus the basic features of the BBT, or some variation of it, is currently enjoying prevailing support in the scientific community.

However, this does not mean that the BBT is by any means the "last word." There are several phenomena in the universe that the standard or "classic" BBT cannot explain. One is the nonexistence of magnetic monopoles,[3] or single magnetic charges.† Another is the extremely delicate balance the universe exhibits between continuing to expand forever and reversing the expansion at some time and ultimately collapsing (more will be said of this later).[4] A third is the lack of an explanation for the average homogeneity of the universe. That is, while there are large clusters of matter in some places and large gaps in others, the distribution of matter in the universe as a whole is essentially

---

*More recent studies have shown that there are some small variations from this spectrum at the higher frequencies, but a definitive interpretation of these results has not yet been formulated. (S. M. Zipoy, Department of Astronomy, University of Maryland, 1983, private communication.)

†It will be remembered that magnetism as we know it arises only from electric currents or from magnetic dipoles (magnets with north and south poles). This is to be contrasted with electricity, which can exist as isolated charges of one polarity. However, there is evidence that one monopole may have been observed in 1982, but this needs substantiation. (A. L. Cabrera, cited in *Physics Today*, June, 1982, p. 17.)

uniform on average. The so-called "inflationary universe" theories, which envision an extremely rapid expansion (much more so than the BBT) of the universe in its initial moments,[5] are examples of models which overcome these objections and are receiving intense attention as of this writing.

The Early Universe and Elementary Particle Physics

The problem concerning the magnetic monopole mentioned above is one indication of how the study of elementary particles comes into play in any modern theory of the universe. But as suggested at the beginning of this chapter, the collaboration between cosmology and elementary particle physics has been particularly productive in addressing events in the very early history of the universe.[6] Through the knowledge gained about elementary particles and their interactions by using very high energy particle accelerators (energies as high as 1000 GeV*), theoretical physicists are able to extrapolate back in time about 18 billion years and reconstruct events as close as 10^{-43} sec to time zero!†

Thus our understanding of the universe is probing backward to an incredibly short time after creation. Theoreticians seem to be able to press their theories ever closer to this limit but never quite reach it. It is as if there is an impenetrable wall which cannot be touched and which defies scientific explanation.[7] Accordingly, physics may be concluding that there was an instant of creation. Conclusions bearing similarity to this, arrived at by spiritual insight, have been expressed in many religious traditions for thousands of years. A fuller discussion of this fascinating subject must wait until Part III.

Let us undertake a survey of the most significant details of the very early history of the universe by taking a brief journey back in time to 10^{-43} sec after time zero. Since the concern here is primarily with events near the time of creation, we pass back

---

*A GeV is one billion electron volts (eV). An eV is the energy gained by an electron when accelerated through one volt.

†Recall that 10^{-43} is a fraction with 1 in the numerator and 1 with 43 zeros behind it in the denominator.

beyond the evolution of life and man, the formation of the earth 4.5 billion years ago, the generation of the stars and galaxies, to 100,000 years (10^5 years) after the Big Bang.

It was at about this time that the universe had cooled down to a state where the formation of stars and galaxies was possible. From the point of view of physics, this is the epoch in which the universe is existing today. Previous to this time, due to the high temperature of the universe, it was not possible for atoms to stay together. The electrons that ordinarily surround a nucleus to form an atom were torn away by the continuous chaotic bombardment characteristic of particles at high temperature. It was after $T_0 + 10^5$ years (T_0 denotes the instant of creation) when atom formation was possible, that much of the primordial background radiation detected by Penzias and Wilson originated. This radiation or light was emitted largely as a result of electrons within the atom giving up radiative energy as they dropped to lower energy states of the atoms in the cooling process. However, previous to $T_0 + 10^5$ light was also being produced from nonatomic processes.

It is conceptually useful to look upon this cooling as a "freezing out" of the nuclei and electrons to form atoms,[8] very much the same as the molecules of water have their movements slowed enough by cooling to form crystals of ice. Indeed in proceeding back to $T_0 + 10^{-43}$ sec, several successive "freezings" occur, each of which defines an epoch of progressively shorter duration.

The underlying mechanisms involved in one way or another in most of these freezings depend on the four forces (or what physicists call interactions) presently known to exist in nature. Perhaps the most familiar is gravity. Another is the electromagnetic force which keeps electrons bound to nuclei to form atoms, and atoms in turn bound to other atoms to form molecules. It is this force that is basically responsible for all chemical and biological reactions and thus is what keeps animals and humans together and operating. The third of the four forces is the strong or nuclear force which holds protons and neutrons together in a nucleus. The last is known as the weak force and comes into play, for example, in types of radioactive decay that involve the emission of electrons and neutrinos (members of a generic particle group known as leptons).

Gravity is by a large margin the least powerful of the four forces. As Trefil puts it, "An ordinary magnet can lift a nail, even though the entire Earth is on the other side exerting a gravitational attraction."[9] The weak force is next, but it is about 1000 times weaker than the electromagnetic, which in turn is roughly 100 times weaker than the strong or nuclear force.

Theoreticians have shown that as the temperature and energy are increased the nature of these forces changes, so that one can unite with another. That is, they can then be described by a common mathematical formalism. Thus going back in time to higher and higher temperatures, while particles disintegrate into smaller subparticles, the forces governing the particles, in contrast, tend to unite.

This will be illustrated in somewhat more detail as we resume the journey back in time by picking up at $T_0 + 10^5$ years. Recall it was after this time that atoms were able to "freeze out" from the high temperature "soup" of nuclei and electrons. But the nuclei in this "soup" were the result of an earlier "freezing" at about $T_0 + 3$ min from a higher temperature soup of protons and neutrons. They were in turn frozen out of a still higher temperature soup of leptons (electrons and neutrinos, etc.) and quarks at $T_0 + 10^{-6}$ sec. It is at this point proceeding back in time and to higher temperatures that the electromagnetic and weak interactions unite. Finally between $T_0 + 10^{-35}$ sec and $T_0 + 10^{-43}$ sec the electromagnetic, weak, and strong forces are all three united. In this regime temperatures corresponding to energies of around 10^{15} GeV are reached, far beyond the energies achievable with manmade particle accelerators (about 10^3 GeV). In effect the universe must now become our laboratory. Theories that embrace the three forces as aspects of a single high-energy interaction have popularly become known as Grand Unified Theories, or GUT.

Theoretical physicists are presently working on the ultimate step which will incorporate gravity along with the other three forces into a supreme mathematical unity. Such theoretical frameworks are often called Super-GUT or Supergravity theories. Some versions of Supergravity require *eleven space-time dimensions,* namely the usual four plus seven additional dimensions characterizing "compactified" or unobservably small, compressed regions of the physical continuum, (perhaps similar to Wheeler's "quantum foam" mentioned in Chapter 2). (See D.Z. Freedman and P. van Nieuwenhuizen, "The Hidden Dimensions of Space Time," in *Scientific American,* Vol. 252, March, 1985, p. 75.) In this connection it is interesting to mention that after his work on relativity and quantum theory, Einstein spent much of the rest of his life on unsuccessful attempts to unify the electromagnetic and gravitational theories.

It is in the regimes mentioned above where the strong and weak forces are changing that the long-standing physical principle of time reversibility begins to break down. This makes possible ultimately the formation of more matter than antimatter, which is apparently the way things are in the universe today. The fundamental relation that this fact had to the arrow of time was pointed out in the last chapter.

All our usually acceptable concepts of time are challenged as T_0 is approached. In this regime of prodigiously high temperatures, enormous compression, and altered laws of force, it would be surprising if our fundamental notions of time and space might not need some drastic rethinking. For example, as we will soon see, Wheeler maintains that time did not even exist antecedent to T_0. On the other hand Schramm speculates that prior to $T_0 + 10^{-43}$ sec there may be no T_0, and an infinite stretch of time may have existed.[10] In any case it is evident that some kind of apparently impenetrable barrier is being reached, but the challenge this barrier presents is made the more fascinating by the simplicity and beauty that characterizes the universe's description as we approach it.[11] More will be said of this when we discuss the thought of some individual cosmologists on this as well as other time-related aspects of the universe.

The Universe: Open or Closed?

One of the reasons for such intense study of the conditions prevailing in the early universe is that more accurate knowledge of these initial conditions helps physicists formulate reasonable extrapolations in the other time direction, the future of the universe. At the heart of these estimates about the future is the burning question as to whether the universe is open or closed. That is, will it continue to expand, as it is now, indefinitely or forever, in which case it would be called an "open universe." Or, is there enough mass in the universe so that at some time the gravitational attraction will cause the expansion to reverse with the subsequent contraction leading ultimately to a "big crunch," as Wheeler puts it?[12] This is an abbreviated scenario for a "closed universe." Here it might be pointed out that regardless of whether it is open or closed, considerable expansion is necessary in order to prevent gravity from very quickly collapsing the universe.

Theoretical cosmologists have calculated the value of the critical density (i.e., critical mass per unit volume) of the universe, above which it is closed and below which it is open. The luminous matter in the observable universe, that is, light-producing stars, galaxies, etc., accounts for only about five percent of the critical density. Thus superficially it might seem that the universe is clearly open. However, there is reason to believe that there is a considerable amount of nonluminous matter in the universe, perhaps even enough to determine that it is closed, or if not, probably enough to bring the density close to the critical point.

One of the simplest sources of nonluminous material is cosmic dust thought to be distributed near the periphery of galaxies.[13] Another possibility now under intense study is that the neutrino, thus far thought to have zero mass, might have a small mass. Using Einstein's famous equation $E = mc^2$ which relates mass and energy, it can be shown that a mass equivalent in energy to only a few eV* would suffice for the universe to reach critical density. There are theoretical reasons for believing that at least some types of neutrinos may have nonzero mass. At present many extremely sensitive experiments are being conducted to test these hypotheses. A third interesting possibility is black holes, which are such enormously intense accumulations of mass that not even light can escape their gravitational pull, hence the name black hole. Since they emit no light and thus are not observable by direct means, there may be many which are contributing unrecorded mass to the universe.

---

* Definition of eV was given at the beginning of the third section of this chapter.

The Open Universe: Endless Time

Since the universe is estimated to be close to the critical density, it is apparent that the issue of whether it is open or closed may not be settled for some time. The fact that it could go either way has generated fascinating projections and estimates about the scenarios of the two alternatives.

In the case of the open universe reasonable predictions based on known cosmological data indicate that the stars will exhaust their nuclear fuel and stop shining at about $T_0 + 10^{14}$ yrs (remember the present time is about $T_0 + 1.8 \times 10^{10}$ yrs).[14] Indeed long before this the sun will have undergone an initial phase in the use of its nuclear fuel, and will swell into what is known as a Red Giant, (a very large star with relatively low surface temperature), swallowing the earth about five billion years from now. Although this is not much for us now to worry about, hopefully our progeny will find somewhere else to go. In any event at 10^{17} yrs the stars will have lost their planets by interstellar collisions, which will be preliminary to the escape of some stars by collision from the galaxies and the collapse of the galaxies to black holes by 10^{18} yrs.

The next step in the open universe story depends on the validity of the grand unified theories, GUT, mentioned earlier. In part because the original GUT predicted too short a life time for the proton, the theories have undergone revisions and now predict a half-life for the proton of between 10^{30} and 10^{33} years. That is, one half of the protons now present will have ultimately decayed into positrons, electrons, neutrinos, and photons in that time. Because the half-life is so long, decay events are extremely rare and involve elaborate and sensitive detection equipment. But the watch for a genuine proton decay is now being conducted by several groups of experimental physicists, with present results indicating that the half-life is greater than 10^{32} yrs. Therefore the current version of GUT appears to be in agreement with experiment thus far, but the situation could change when an accurate value is available for the proton half-life, assuming it can be observed to decay at all.

Though it now appears to be uncertain as to whether proton decay is the next step in the open universe scenario, nevertheless the last stage is the decay of the black holes. Theoretical cosmologist S. W. Hawking has shown through the use of the Heisenberg Uncertainty Principle that such decay should be expected. Thus the universe would ultimately end as a diffuse gas of electrons, positrons, neutrinos, and photons cooling to even lower temperatures as the expansion continues and the cosmological clock ticks on.

There are other theories which support the clock's ticking on in an open universe. Among these is the idea that the gravitational force very gradually weakens as the expansion progresses so that the point may never be reached where it reverses. This concept is one that, until his recent death, was long espoused by Nobel Prize winner, P. A. M. Dirac,[15] who was the first to propose the existence of the positron.

In addition to reasonable scientific extrapolations as to the future of the universe based on known data, some cosmologists have attempted to estimate how some form of intelligent life might be propagated into the indefinite future. Although by their nature such efforts are speculative, they are based on comprehensive thought and serious calculations. Among the most fascinating scenarios of our future in an open universe is that suggested by Freeman Dyson.[16] He prefers the prospect of the progressive cooling of the open universe rather than the torrid implosion ending the closed universe because, as he puts it, he would rather "freeze than fry."

Probably the most fundamental of the assumptions that Dyson makes in his calculations is that consciousness is based on structure, not on matter. For example, the similarity between the human brain and a computer lies in their structure, i.e., the peculiar arrangement and interlinking of the components; it does not depend so much on the material the components are made of. In the brain the arrangement and interaction are of the carbon-based molecules; in the computer they are of the silicon-based minicomponents. Thus his calculations envision our ultimate evolution into a more lasting form of embodiment than blood, flesh, and bone. In fact, Dyson visualizes a kind of "sentient black cloud" that has all of the computer memory, thought capacity, and communication ability of the human brain without its material frailties, that could more easily adapt to the ever colder atmosphere. By undergoing carefully chosen periodic hibernations, it would use less and less energy to operate and to communicate with other such beings. Because of the ever-decreasing use of energy, such intelligent beings could exist into the indefinite future, according to Dyson.

A somewhat similar suggestion is made by Jastrow who calls contemporary humans "living fossils."[17] He portrays a progressive development and interaction between us and the computers we build. As the sophistication of the computers increases and our interaction with them becomes more intimate, they will ultimately be able to exist and reproduce on their own, forming a more lasting silicone-based society as we carbon-based humans gradually die out.

The Closed Universe: Will Time End?

As indicated earlier, if the average density of the universe is greater than the critical density, then the mutual gravitational pull will ultimately cause the expansion to cease and a contraction will set in. The closer the average density of the universe is to the critical density, the longer will be its lifetime. Since apparently the average density is probably quite close to the critical density, the present expansion could continue for a very long time, in fact so long that many of the events described for the open universe will also occur.

Thus the universe at its maximum expansion before beginning to contract could well be made up of dead stars that have radiated away all their excess energy, massive black holes which are the result of galaxies' gravitational collapse, and a "gas" of photons and neutrinos.[18] As the contraction commences and proceeds, some of the dead stars will begin to heat up by the interaction of progressively more energetic photons. Remember that the Doppler effect indicates there will be a blue shift to shorter wavelengths, higher frequencies, and greater energies. Such stars under these conditions will gradually evaporate, the resulting material being sucked up by giant galactic black holes. Other dead stars may be consumed by the black holes directly. In the advanced stages of the contraction more and more black holes will coalesce until the universe becomes one incredibly massive black hole. The "big crunch" will have occurred and time will end.

But will it? One of the most interesting features of the closed universe concept is the implied possibility, due to some unknown mechanism, for another explosion, thus commencing another cycle of expansion and contraction. Although admittedly not knowing the process by which the universe could "bounce" back, some cosmologists[19] have made calculated estimates of such cyclic behavior. One clear result of these estimates is that each cycle will be longer than the last by at least a factor of two, but this factor could be very much larger. The relationship between such a cyclic picture and that envisioned in some of the cosmologies of the Hindu religious tradition will be discussed in later chapters.

The Anthropic Principle

Before discussing some suggestive implications of a closed universe, especially the thought of John Wheeler, it will be useful to describe a very relevant concept known as the Anthropic Principle. As the name may suggest, the principle has to do with the relationship between man, or some observer, and the universe, and in many instances has been found useful to cosmologists. In essence the principle makes use of the simple fact that we humans have evolved. That is, this fact of our evolution is employed to place restrictions on the range of possible models of the universe and its beginnings that can be considered. In other words, if the initial conditions of the Big Bang were a little different, with a concomitant small change in the physical constants and the physical laws of which they are a part, we would not be here. Thus the fact that we are here gives cosmologists guidelines as to what the behavior of the early universe must have been.

As will be seen later, the Anthropic Principle, when used in its extreme interpretation, places man in a rather central position in the universe. This is in contrast to what is known as the Copernican Principle which states that there are no privileged or central positions anywhere in the universe.* A more modified position is expressed by Brandon Carter[20] who states: "Although our situation is not necessarily central, it is inevitably privileged to some extent." The universe apparently had to be that rather unique universe, out of all that might reasonably be formulated, that made possible the evolution of its own observer.

It seems remarkable that the constants of physical nature have just the right values to make it possible for us to be here. If a constant known as the fine structure constant were just a little different,† the convection currents of stellar material around a star would not be able to develop into a planetary system, and we would not be here. If the strong force keeping neutrons and protons together were just a little weaker, hydrogen would be the only nucleus and we would not be here. The calculations of Collins and Hawking show that the expansion velocity of the universe is at that critical value where if it were greater the universe is open, if less, closed.[21] Is the delicate balance of the universe being at this critical value also related to our being here?

In any case there seems to be a systematic succession of delicate physical balances, predetermined at the Big Bang, which have made our evolution possible. This is regarded by many serious thinkers as evidence for a philosophical claim of some degree of centrality for the human observer. Such a claim also has some clear religious implications, some of which are time related, to be developed in Part III.

The Universe of John Wheeler

While he does not deal in religious questions, in his thought John Wheeler places the human observer in a rather vital role in the universe, a universe which he seems most often to assume is closed. In developing his ideas he makes use of a strong interpretation of the Anthropic Principle in which he suggests "the

---

*Recall that it was Copernicus who showed that the earth revolved around the sun instead of the reverse, as was strongly believed at the time.

†The fine structure constant is the ratio of the square of the electron charge to the product of Planck's constant (see Chapter 2) and the velocity of light.

universe could not have come into being unless it were guaranteed in advance to be able to give rise to life at some point in its history-to-be."[22] The universe had just the right size and balance between mass and expansion velocity along with just the right physical constants to make possible the 18 billion year "cooking" time of the soup of nuclei and electrons in the correct proportions to make stars, planets, and man. To quote Wheeler: "Why then is the universe as big as it is? Because we are here."[23]

He then takes seriously the question posed by Dicke who asks in effect, "what possible sense it could make to speak of 'the universe' unless there was somebody around to be aware of it."[24] Therefore, Wheeler sees a very intimate interaction between the presence of man, the observer, and the universe; and as explained in Chapter 2 when discussing quantum observer theory, he prefers the term "participator" to "observer." It will be remembered there how he suggested that, *in a sense*, depending on the measuring apparatus used and its setting, the measurer in the present influences the nature of a physical event which occurred billions of years ago, for example, the background radiation from the Big Bang.

Wheeler then, drawing on the conceptual connection in quantum theory between the observer and physical phenomenon observed, is prompted to ask if "the universe, through some mysterious coupling of future with past, required the future observer to empower the past genesis?" Somehow "observership allows and enforces a transcendence of the usual order of time" so that the "observer is as essential to the creation of the universe as the universe is to the creation of the observer."[25]

Other insights about our knowledge of the physical world are used by Wheeler to supplement and support his ideas. He goes into some detail to consider the progressive superseding of one physical law by its more sophisticated successor over the last several hundred years of the history of physics. The laws are generally expressed as symmetry or conservation principles. His point is that starting with Archimedes' principle concerning the density of matter, each law has hidden from view the more sophisticated physics that underpins the law. A deeper law is then formulated with the discovery of more complicated physical phenomena. This has gone on in a more or less step-wise process. However, realizing that in a closed universe the "big crunch" will be the end of all laws, Wheeler senses there must be an end to the process.* He speculates that there must be a closed circuit of the interdependencies in these laws, and that the circuit is closed by the observer himself. To quote: "Could it be that the observership of quantum mechanics is the ultimate underpinning of the laws of physics—and therefore of the laws of time and space?"[26] In fact he goes so far as to suggest that the greatest challenge to the physicist is to attempt to derive the quantum principle from the concept that genesis is dependent on observership.

These and related considerations have caused Wheeler to conclude that time as a fundamental concept may be in trouble.[27] He bases this in large part on some of the phenomena already discussed and maintains that there was no time before the Big Bang, and there will be none after the big crunch, if the universe is closed† (unless the universe is cyclic). He also bases his conclusion on what was learned in Chapter 2, namely that at infinitesimal distances and times beyond the limits of measurability governed by the Heisenberg Uncertainty Principle, time has no meaning.

Wheeler therefore feels that time is a derived quantity, derived from a more fundamental yet undiscovered physical principle or law. He suggests that the way leading to this law is via the quantum theory.[28] One property of this law is that all our known physical laws change, e.g., as they do when we look back at the Big Bang. This aspect of the new undiscovered physical principle he calls the Law of Mutability: there is no law that is immutable. He maintains a way must be found to derive such mutable laws, along with mutable time, from almost nothing.[29] Actually efforts along these lines are now being pursued by theoretical physicists.[30] While normally it is assumed that the vacuum has zero energy, viable theories in elementary particle physics have

---

*This end might bear some similarity to the changes in the physical laws already discussed as the Big Bang is approached going back in time.

†In fact since "before" and "after" imply time, they should not be used here. They are used for lack of better expression.

been formulated which assume that the vacuum (Wheeler's quantum foam) does have some energy per unit volume, as discussed in Chapter 2. This makes it possible through Einstein's $E = mc^2$ for something to emerge from "nothing," at least for a very short while consistent with the Heisenberg Principle. This concept bears some relation to many religiously based creation myths and to the assertion of St. Thomas Aquinas that God created the universe "ex nihilo," out of nothing. The ramifications and relationship of these views and their bearing on time will be discussed in Part III.

Summary

About 18 billion years ago the universe was born with a gigantic explosion, the results of which are still observable. Among these data are: 1) the observation that light received from receding stars and galaxies is shifted toward the red end of the light spectrum as determined by the Doppler effect; 2) primordial low energy radiation emitted in the early history of the explosion is still detectable. These phenomena among others support the general features of the Big Bang Theory, BBT, which now appears to need further refinement because there are phenomena (e.g. magnetic monopoles) which it cannot explain.

The early history of the universe was treated in terms of "epochs" wherein each was characterized by a "soup" of particles "frozen out" from the hotter "soup" of the previous epoch. This scenario arises from predictions of the Grand Unified Theory, GUT, in conjunction with the BBT. GUT in addition predicts a progressive unification of the forces of nature proceeding back in time to the instant of the Big Bang. Though GUT is probably still subject to revisions with respect to some of its predictions, some version of it may some day prove to be valid.

The subject of whether the universe is open or closed was discussed, with an emphasis on the fact that it appears to be very near the critical point between the two possibilities. Scenarios for the two were described, along with speculations about the survival of some form of intelligence in the case of the open universe. The treatment of the Anthropic Principle, which requires that any theory of the universe must explain our presence in it, was followed by the thought of John Wheeler, which is based primarily on the assumption of a closed universe.

Although some such as Schramm might not agree,[31] others (e.g. Wheeler) feel that time did not exist prior to the Big Bang, nor will it in a closed universe after the "big crunch." The similarity between such a view and the thought of St. Augustine will be discussed in Chapter 9 and in Part III.

Since elementary particle physics tells us that physical laws change as the moment of the Big Bang is approached, this consideration among others prompts Wheeler to conclude that time is not a truly fundamental quantity but one derived from a yet-to-be-discovered physical principle. According to him, this principle will entail the derivation of almost all known and presently valid physical laws, as well as time, from *almost* nothing. It may also, as he suggests, have to include man the observer-participator in a much more intimate way than have any previous laws.

Finally, the universe is looked upon by many as a kind of cosmic clock ticking off the increments that move the cosmological arrow of time. As mentioned earlier, the duration of time according to this clock has been about 18 billion years. As we will see later, this duration is comparable in scale to those envisioned in some of the cosmologies in the Hindu religious traditions.

II

Some Religious Views of Time

Here we must be prepared for some kind of "cultural shock" as we pass to the other extreme of the intellectual spectrum. In this part we examine some concepts of time in the Hindu, Buddhist, Taoist, and Judeo-Christian religious traditions. In each chapter relevant background material is presented, followed by the views of time from these traditions that bear meaningful comparison to the physical concepts discussed in Part I.

5

Hindu Cosmic Cycles:
Manifestations of the Timeless Brahman

Hinduism is the oldest living major religion in the world, with roots that date back to ancient times in India and Pakistan. In part because of its age, in part because it is not based upon the teachings of one man with a centralized clergy, and in part because of the prolific intellect, intuition, and imagination of its many leaders, it is a religion of unparalleled richness and diversity. Indeed the diversity is so great, that many of the beliefs and philosophies actually contradict one another. Thus, to distill from such a diverse and enormous matrix of philosophies something like a common thread on any subject is not a simple matter. This is true in particular of the two closely interrelated facets of Hindu tradition we now wish to examine, time and cosmology.

The Development of the Hindu Religious Tradition

For a richer understanding of the concepts involved, it will be helpful to give a brief outline of the depth and scope of the development of Hinduism. Its origins date back to about 1500 B.C. when Aryan tribes from the steppes of Eastern Europe invaded and settled in what is today Pakistan and India. The declining, indigenous Indic civilization was gradually assimilated by the Aryans, whose religious beliefs were based on four collections of hymns called *Vedas*.*

The earliest of these, the *Rig Veda,* was the scriptural begin-

---
**Veda* means "knowledge" or "body of knowledge."

ning for a religion in whose early history rituals of sacrifice played an increasingly central role. Such sacrifice, usually involving fire along with sounds and words (mantras), came to be regarded as a power in itself. The cosmos, the natural world, human society, and sacrifice were all four regarded as "parallel orders of reality of equal antiquity and permanence."[1] But the central, unifying element in this system was the sacrificial ritual.[2] Therefore, by knowing sacrifice, one could not only know the other three orders, considered structurally similar, but also actually know and control the universe. The goals were health and prosperity during life, but the ultimate goal was immortality in heaven, the world of the gods, after death.

However, as time went on a more speculative literature emerged in the later Vedas, such as the *Atharva Veda,* and another body of scripture known as the *Brahmanas.* And in the later Brahmanas doubt began to be expressed about the value of immortality and the use of sacrifice to attain it. Gradually immortality came to be viewed as an endless chain of deaths and rebirths (as it is in the world of nature) and was not felt to be such an attractive prospect. A deeper insight about the nature of man and his place in the world was sought.

In this process the speculative probing increasingly developed a distinction between the main body of ritual texts and the texts that emphasized the meaning of the rituals. Many of the latter were embodied in a third scriptural group called the *Aranyakas,* or "Forest Books," so named because their contents were taught and practiced in forest retreats.[3] However, it was the *Upanishads,** the last group of Vedic literature, which carried Vedic speculation to its deepest level.[4] With the emergence of the *Aranyakas* and the early *Upanishads* the speculative search led to a growing realization of the importance of the individual. The self began to be identified with the Self or Atman, the indestructible spirit of the god Brahman in man. And Brahman began to be looked upon no longer as the power in sacrifice but as the ultimate reality of the universe.

Nevertheless, by about the 6th century B.C. there was a grow-

---

*The Vedic literature including the *Vedas, Brahmanas, Aranyakas,* and *Upanishads* comprised a scriptural group known as the *Sruti* (that which is revealed or heard).

ing disillusion with what Brahmanism (essentially the earlier form of what is now known as Hinduism) had to offer, in part because it provided no spiritual escape from the endless cycle of rebirths and deaths known as the cycle of *samsara*. Furthermore, much of the wisdom of the priests was exchanged among themselves and not available to the people; the personal solutions to the problems of human existence were not being addressed.[5] Consequently new religious movements such as Buddhism and Jainism sprang up. Buddhism, to be discussed in the next chapter, opened its wisdom to the layman. In particular this applied to rulers. The most notable example was the great emperor Ashoka who, overcome with remorse after a bloody conquest consummating his rule over India, converted to Buddhism along with multitudes of followers.[6] This was a great blow to Brahmanism.

However, the religion learned by its mistakes, and by reaching out to the people through adapting to some of their family religious practices, and especially by assimilating yoga, a resurgence of faith and support developed. It was this ability to assimilate, adapt, and synthesize that saved the religion and made future growth possible. This was reflected in the later Upanishadic works. For example, the *Svetasvatara Upanishad,* often distinguished as one of the theistic *Upanishads,* emphasized the personal aspects of Brahman, as opposed to the impersonal, and stressed that salvation was available to anyone through proper meditation.[7] One way that a synthesis of the old and the new was accomplished was by unifying the old Vedic use of sounds and mantras, such as "Om," with the newer popular practice of visualizing or imagining the gods as having some human form as well as human attributes.

One of the most beautifully expressed and fully matured syntheses was effected in the *Bhagavad Gita* which is a part of the *Mahabharata,* probably the world's longest epic poem with some 100,000 verses. The *Gita,* as it is often called, is one of Hinduism's most beloved religious texts.[8] In it the dialogue between the warrior Arjuna and his charioteer and mentor, who is later revealed to be the god Krishna, is used as a literary vehicle to express its message of faith, action, and love. The scene is a battlefield where the kinsmen of a large royal family are lined up in opposition, ready for battle, to decide who gains the family kingdom. Arjuna is suddenly struck with guilt and indecision over the prospect of slaying his kinsmen. Would it not be better for him to be slain instead?

Krishna becomes his teacher, urges him to fight, and gradually persuades him that, though the body may be slain, the inner self is not.[9] One should take action and perform one's duty with faith and without concern for the karmic consequences.* He stresses to Arjuna that it is not the actions themselves that are the problem, but our attachment to the fruits of the actions.[10] In the discourse he first reveals his identity as the god Krishna, but later as the supreme Atman, the Self, the spiritual manifestation of Brahman in man. In an artistic and skillful way, important Vedic and Upanishadic concepts are synthesized with newer concepts stressing the importance of Self. The Vedic notion of sacrifice is woven in by the teaching that one's duty should be regarded as a sacrifice to the gods. The impersonal Brahman, characteristic of many of the *Upanishads,* is unified with a new more personal one, the manifestation of which offers a way to salvation open to all levels of society.[11] In sum, the impending battle pondered over in the *Gita* is a beautiful metaphor addressing the struggle of life, offering the hope of salvation for all through faith, action, and love.

However, the poetic *Epics*—the Mahabharata (which, recall includes the Gita), and another known as the Ramayana—actually reveal only the initial stages of Hinduism's theistic development. The most comprehensive record, carrying this development to its fullest maturity, is found in still another body of literature called the *Puranas,* which, among other things, deal with the creation of the universe and its periodic destruction and rebirth, and provide the principal source for our knowledge of Hindu cosmology.† The *Puranas* to this day are the primary scriptures of theistic Hinduism.

---

*Karma, often called the law of cause and effect, essentially expresses the belief that one's fortune in this or in a future life is affected by one's own action.

†Generally, what is termed a cosmology in a religion is not completely such in the strict physical sense. While it usually does possess some quite specific physical properties, we whall see that it is often also impregnated with a spiritual and/or soteriological (salvational) meaning.

However, for a more detailed understanding of Hindu time concepts, we must look to a final set of texts called the Sutras.* Written after the *Epics,* these works represent the attempts of thinkers to systematize philosophically the views about the vast literature thus far accumulated. These efforts resulted in six principal schools of religious philosophic thought: Nyaya, Vaisesika, Sankhya, Yoga, Purva-Mimamsa, and Vedanta. The six schools can be reduced to three groups in terms of basic philosophic premises. The Nyaya and Vaisesika essentially have a pluralistic view of the universe, the Sankhya and Yoga a dualistic one, and the Purva-Mimamsa and Vedanta a monistic view. In a monistic philosophy the universe is ultimately based on a single irreducible entity; in the dualistic two such entities are involved; in the pluralistic, more than two. The *Sutras,* presenting the views of the various schools, depend largely on logic in contrast to the intuition expressed in the *Upanishads.* It is therefore not surprising that some very specific philosophic views of time and cosmology are set forth in many of these works. Pertinent and selected aspects of these concepts of time will be discussed later in this chapter. It is sufficient here to note that it is the Vedantic school with its offshoots, embodying the views of such great leaders as Shankara, Ramanuja, and Madhva, that now basically speaks for modern Hinduism. However, it is Shankara's Advaita Vedanta that gives the most systematic presentation of Upanishadic insights, and that is the most dominant in Indian philosophy and culture. It is Shankara who most eloquently and beautifully expresses the monism or non-dualism that is at the heart of the mainstream of Hindu religious philosophy.[12] The universe and all life therein are but a transient manifestation of the ineffable One, Brahman.

Time and Cosmology in the Central Hindu Beliefs

With this survey of Hindu history and literature in mind, we can now attempt to extract some basic concepts of time and

---

*The *Sutras, Puranas, Epics,* and another body of texts, called *Dharma Shastras,* devoted largely to moral law and social duty, make up a second scriptural group known as the *Smriti* (That which is remembered).

cosmology by looking at some of the generally accepted central Hindu beliefs. The ultimate deity under which all other deities are subsumed, and from which all derive their power, is Brahman. Brahman is the Absolute, the only true reality: beginningless, endless, changeless, indescribable, and also beyond good and evil, nature, the universe, time, space, and causation. The ultimate individual Self, also beyond the ensnaring concerns of this world, is Atman, which is identified inseparably with Brahman. In other words, Atman is the indestructible spirit of Brahman in man. Beginning in the *Upanishads* and more fully developed in the Vedantic texts is the distinction between two aspects of Brahman, the supreme (Nirguna) Brahman just described, and an inferior (Saguna) Brahman, Isvara, Lord of the knowable universe, with attributes (in contrast to the attributeless Nirguna Brahman), among them, the capacity to be worshipped on a personal basis. The power of Saguna Brahman expressed in the universe is *maya,* variously interpreted as the perceived world or nature *(prakriti),* and considered as basically illusory because of its essential impermanence. The operation of the universe is the result of the action of the three great divinities who are manifestations of Saguna Brahman: Brahma, creator of the universe; Vishnu, its preserver; and Shiva, its destroyer.

The very nature of this trinity implies a cyclical universe that at the end of the periodic cycle is destroyed and absorbed into the indescribable Brahman, then recreated for the next cycle:*

> All beings, O Son of Kunti (Arjuna), enter into My nature at the end of a world cycle, and I send them forth again at the beginning of a new cycle. *(Bhagavad Gita 9.7)*[13]
>
> After having created all the worlds, He, their Protector, takes them back into Himself, at the end of time.† *(Svetasvatara Upanishad, 3.2)*[14]

As generally accepted in the Puranas, the smallest period of the Hindu cosmic cycle is the *yuga,* of which there are four,

---

*Such a concept of world cycles suggests correlation with the notion of the human samsaric cycle of life, death, and rebirth.

†Here the end of time is interpreted by later Hindu scholars as meaning the end of a cycle.

which total to 4,320,000 years. Each is progressively shorter by one quarter, ranging from the krita yuga of 1,728,000 years to the kali yuga of 432,000 years. Also each is characterized by successively decreasing *dharma* or moral order. We are now in the kali yuga whose end will involve chaos, war, and dissolution of all moral standards and cultural traditions. Then a new cycle of four yugas, known as a mahayuga, will commence again and last for another 4,320,000 years. One thousand mahayugas (4,320 million years) is a *kalpa,* one day in the life of Brahma, Creator of the universe. At the end of the day the universe is reabsorbed into Brahman, and the night of Brahma, also one kalpa in length, takes place. At the end of the night Brahma is reborn. The "Brahmic" year consists of 360 days, and Brahma is to live for 100 Brahmic years. This yields a total cycle of 311,040 billion years, after which total dissolution occurs for a Brahmic century, whereupon the vast cycle recommences.[15]

Hinduism is the only major religion which envisions specific time durations of such colossal scope. What is striking here is that these giant cycles involving durations of billions of years represent time scales comparable to the current age of the universe, i.e., 18 billion years.

It is very important at this juncture to emphasize the intimate relationship between such cyclical cosmologies and salvation. It is as if the endless periodicity is presented deliberately as an incredible monotony of continuous deaths and rebirths through which one must suffer if what is known as *moksha* is not achieved. Moksha is release from Maya (the natural world and its ensnarements), from the bonds of karma and the endless cycle of rebirths, to a transcendent state of consciousness wherein one is totally identified with Brahman. This state is usually attained through yogic practices. The path to moksha then presents a much more attractive alternative. As we will see in the next chapter, there is also a strong soteriological (salvational) motivation implicit in the Buddhist cosmologies.

Related to the above-mentioned cyclic cosmology are a variety of creation myths which range from creation out of nothing in the *Rig Veda,* to creation out of a gold and silver cosmic egg in the *Chandogya Upanishad.*[16] However, it should be kept in mind that what is described in early writings as a beginning or creation is usually interpreted in later scriptures as the beginning of the present cycle. At the end of each cycle the universe and all living creatures that have not been liberated from maya and the samsaric cycle return to Brahman, regardless of their spiritual condition. The cycles can be more accurately regarded as alternating periods of manifestation and nonmanifestation of Brahman. This deep sense of cyclicality was developed from the earliest times when the Vedic altar was considered to be time itself with 360 bricks for the days and 360 stones for the nights.[17]

The importance that time itself is accorded is apparent in the extensive attention that is devoted to its scrupulous examination in the six philosophic schools mentioned earlier. A complete discussion of this very intricate subject is out of place here and can be found in other sources.[18] However, a very brief sampling of some of the concepts will illustrate the enormous variety found among the six schools. In the Nyaya and Vaisesika schools, which are pluralistic, the multifarious entities experienced in the universe are divided into nine classes, of which time is one: earth, water, fire, air, *akasa* (ether), time, space, self, *manas* (mind).[19] To the Purva-Mimamsa school all perceptual experience, through whatever sense it is acquired, includes a reference to time. However, time cannot be apprehended by itself but only in conjunction with some other object that affects our senses.[20]

Even within some schools the views of time may vary. For example, in the Vedantic school Shankara did not regard time as real, being only a part of the impermanent universe. On the other hand the Vedantist, Ramanuja, viewed time as real but not outside the sole and ultimate reality, Brahman.* Thus it does not subsist by itself, as in the case of the Nyaya and Vaisesika Schools where it is one of nine realities, nor is it a phase of prakriti (natural universe). Therefore time and prakriti are on equal footing and are prior to space.[21] The third great Vedantist, Madhva, in his basic philosophy saw two kinds of time and space, relative and absolute. In the former, space and time evolve from matter; in the latter, they are eternal. Therefore the

---

*Ramanuja, who espoused a qualified form of monism, is generally considered second in stature to Shankara in the Vedanta school.

perceptions of time range from the pluralistic Nyaya and Vaisesika view of time as one of several absolute realities to the Vedantic view of Shankara wherein time is but a temporary manifestation of Brahman and thus without absolute reality.

Since, as indicated earlier, it is the monistic insights of Shankara (comprising what is known as the Advaita Vedanta) that represent the mainstream of modern Hindu belief, let us further elaborate on his concept of time. He sees Brahman as beyond or outside of time. Hence the notion of timelessness is inescapably apparent here. Indeed one can find no more powerful and awesome expression of the concept of timelessness than in the Advaita Vedantic literature. Since Brahman is the only absolute Reality, then timelessness associated with Brahman is endowed with a reality not possessed by time. That is, only things are real which neither change nor cease to exist.[22] The natural universe along with time are only inconstant manifestations of Brahman; they emerge from Brahman and return thereto. The essence of these thoughts are succinctly summarized in Shankara's words: "Brahman is real, the universe is unreal."[23] This implies that time, since it is in the universe, is also unreal, and only timelessness is absolutely real.

One then has the picture of the living, natural universe including time undergoing unceasing cycles of creation, life, and destruction. But beyond and outside of this tumultuous activity is Brahman in timeless, transcendent, and imperturbable majesty. Therefore endless cyclical time superimposed on a background of timelessness figures in a beautifully coherent picture of the cosmos.

It is these two features, timelessness and cyclical time, that I wish to highlight in this chapter, and that are principal characteristics in the mainstream of Hindu thought regarding time. They represent the distillation of millennia of religious growth and synthesis. It was a maturing process that has manifested a profound reverence and respect for time since the Vedic era, as is apparent in the following passage from the *Atharva Veda* with which I close the chapter:

> Time generated yonder sky, time also the earths: What is and what is to be stands out set forth by time.

Time created the earth; in time burns the sun; in time are all the existences; in time the eye looks abroad.
In time is mind, in time is breath, in time is name collected; by time, when arrived, all these creatures are glad." (Atharva Veda 19.53)[24]

Summary

Undoubtedly the most outstanding feature of the Hindu tradition with respect to time is the cyclical cosmology wherein the cycles can be regarded as alternate periods of manifestation and nonmanifestation of Brahman. These tremendous cycles involve time durations of billions of years, durations which are comparable in scale to the current estimates of the age of the universe. This interesting comparison between spiritual insight and physical measurement will be elaborated upon in Part III.

The other very important time-related feature in the Hindu tradition is the concurrent profound apprehension of the notion of timelessness. Although Brahman is considered to be indescribable, one has to use some words in order to at least provide a sense of what the true Reality might be, and *timeless* is certainly one of the principal words used in any such attempt. In no other religious tradition is the aspect of timelessness so deeply imbedded and so supremely expressed. This too will be discussed in Part III where comparison is made with notions of timelessness drawn from results of physical cosmology.

6

Buddhism, Nirvana, and Time

Unlike the Hindu religion and philosophy, Buddhism is based on the teachings of one man, Gautama Buddha. It is generally accepted that his biography has both legendary as well as true historical components. His birth is estimated at around 560 B.C., near the Himalayan foothills. Growing up in the highly protected environment provided by his father, the local prince, he experienced with deep sensitivity and compassion the misery prevalent in the outside world on the few occasions when he was exposed to it. As a young man, he left wife and child to seek an answer to such suffering. After working with two teachers, he ultimately arrived at his own insights and enlightenment and launched his ministry.[1]

The Buddha's prime motive in undertaking this ministry was compassion supported by his willingness to share his enlightenment. He taught the path of the "Middle Way" between self-denial and asceticism on the one hand and sensual indulgence on the other. In the opinion of one scholar, R. H. Robinson, his achievement in releasing ethics from ritual (e.g., the Vedic sacrificial ritual discussed in Chapter 5) was as momentous as Paul's freeing Christianity from the Hebrew Torah.[2]

Since he was brought up in the Brahmanic tradition, the Buddha's ministry thus constituted one of several splits with orthodox Brahmanism that occurred during a general time period spanning much of the 5th and 6th centuries B.C. The foundation of the Buddha's doctrine is the Four Noble Truths: suffering, cause of suffering, cessation of suffering, way leading to cessation of suffering. The last is known as the Eight-Fold Path, involving eight practices among which are: right views or understanding, right intention or thought, right speech, and right action.

Probably the Buddha's most important insight was the perception of the painful drama of continual transmigration (birth, death, rebirth, etc.) as a stream of twelve interdependent elements or key human characteristics. This view is expressed in his law of twelve elements of Conditioned Coproduction (or Dependent Origination), often called the Wheel of Samsara because each of the elements arises from or is conditioned by the others. Among the twelve elements are: aging and death, birth, becoming, appropriation or grasping, craving or desire, and feeling or sensation. Aging and death depend on birth, birth depends on becoming, which depends on appropriation, which in turn depends on craving, etc. However, it should be kept in mind that these elements do not necessarily constitute a chain of cause and effect; their operation is more one of mutual or reciprocal interaction. In any case it is the goal of every serious Buddhist to extricate himself from the samsaric wheel of suffering, transcend it, and achieve the absolute and the unconditioned, i.e., nirvana.

Nirvana

In the words of the well-known scholar of Buddhism, Edward Conze, "Nirvana is the raison d'être of Buddhism, and its ultimate justification."[3] Human words are inadequate to describe the real nature of the Absolute Truth or Ultimate Reality which is nirvana. However, because something is inexpressible does not mean it is nothing.[4] We are thus left with words, which are the only tools available to us in this and other such cases, e.g., descriptions of Brahman in Hinduism or of the experience of the Christian mystics in their quest for unity with God.

With this in mind let us make an attempt by noting that nirvana is not conditioned or dependent on anything, any thought, feeling, or action, worldly or unworldly, or any of the twelve elements in the samsaric wheel.* It is only attainable once all

---

*Conditioned and *unconditioned* are important classifications assigned to various essential properties or entities in the world by the Buddhists. When something is conditioned it means that it is dependent on, or influenced by, something else; it is not eternal. Nirvana is unconditioned.

conditioned or dependent existence, becoming, desire—in particular desire for nirvana itself—has been given up. All conceptions of it are misconceptions. It cannot even be achieved by a desire for death, which is just another desire. This means that nirvana can be attained in this life. Though the groundwork for its attainment may be prepared for by following the Eight-Fold Path, this does not guarantee the result. It cannot be wished, or willed—it happens. In its final achievement all effort and striving must cease and even the Eight-Fold Path in its turn is abandoned, like a raft on reaching shore.

As we will see shortly, in many Buddhist schools of thought time is a conditioned entity, dependent on the living world; but in all schools, nirvana is unconditioned. Thus in nirvana time has been left behind; timelessness prevails.

Concept of Non-Self

While Buddhism possesses some general similarities to the Hindu beliefs and practices (e.g., the idea of reaching Brahman [moksha] bears some rough similarities to nirvana), the Buddha departed from Brahmanistic orthodoxy by not asserting an essential or ultimate reality in the universe, or dealing with a concept of first causes, or world-creation theories. The Buddha never actually repudiated the Brahman of Vedic teaching. He tended to be pragmatically disposed to living in the here and now and not given to metaphysical discourses.[5]

Perhaps the most significant departure was his position that the self (soul) does not exist. That is, there is no real essence by means of which man can be distinguished from anything else. Man is simply composed of five aggregates (known as *skandhas*): matter, sensations, perceptions, impulses and volitions, and consciousness. This belief, often termed *anatman,* is directly opposed to the Hindu concept of Atman, Self, the spirit of Brahman in man. The doctrine of anatman is one of Buddhism's most distinguishing features and places the religion in a unique position among the world's major religions.

Buddhist Sects

Within about 200 years after the Buddha's death (about 480 B.C.), as Buddhism spread, schisms began to occur within the religion, ultimately leading to two main divisions: the Hinayana, holding to the earliest traditions, and the Mahayana, which developed in the period between about 100 B.C. to 100 A.D. The Hinayana, generally considered more traditionally orthodox, tended to be more dominated by the priesthood, so that it was usually only a priest who could achieve nirvana. The Mahayana tended to be more democratic; the path to nirvana was open to all, and the salvation of one's neighbor and universal enlightenment were stressed.

Both of these main divisions fragmented into many sects as the religion spread to Ceylon, Southeast Asia, Tibet, China, and Japan, ultimately leaving India almost entirely. One of the leading schools in the Hinayana for which time was very important was the Sarvastivadin. The Sarvastivadins believed that everything is real and exists, in particular, the past, present, and future. This ran counter to many other Buddhist schools of thought which held that only the present exists.

Another especially interesting Hinayana sect was the Pudgalavadin which, contrary to the mainstream of Buddhist thought, believed in the existence of some form of self or soul, called the *Pudgala*. The Pudgalavadins claim that there is a Pudgala in addition to the five skandhas listed above, but it is not the same as the skandhas. Neither is it different from them, in the sense that it furnishes some kind of essence for the aggregate of the five. The Pudgala's relation to the skandhas might be considered analogous to the relation between fire and fuel. Thus the concept is rather subtle and nebulous, so that the Pudgala is not the self in the conventional Western sense.

Two of the leading schools of the Mahayana are the Madhyamika and the Yogacara. The latter basically holds to the belief that only the mind is real, while the former, sometimes known as the school of the Middle Way, criticizes essentially all religious philosophic positions. As such it tends to stand between the Yogacara (only mind is real) and the Sarvastivadins (all is real). Except for the Buddha himself, Nagarjuna, founder of the Madhyamikas, is considered by many as the greatest of the Buddhist leaders.

Cosmology and Time

The Mahayana generally have a flair for the mysterious, miraculous, and imaginative. This is particularly apparent as a characteristic of their religious cosmologies. Identifying the attainment of nirvana with Buddhahood, some sects, such as the Pureland sect, visualize endless worlds throughout space inhabited by Buddhas. This accommodates the belief that nirvana or Buddhahood is available to all. Therefore their cosmology places an emphasis on space. In contrast, in the Hinayana some sects hold to the belief that there was a succession of some seven Buddhas, the seventh being the historical Buddha. Thus their cosmologies tend to be dominated by time.

Although discussion of cosmologies may appear contradictory to the earlier remark concerning the Buddha's apparent aversion to indulge in such subjects, they nevertheless became quite significant in many sects, particularly those of the Mahayana, e.g., the Pureland sect. This is in part because to many of the Buddhist teachers the cosmologies represented or implied the drama of salvation, or path to salvation (nirvana),[6] which, it will be recalled, is also a strong component in the Hindu cosmologies.

Buddhist cosmologies generally have kalpas as their unit of measure for the cosmic cycles, and these are roughly similar to those in the Hindu systems. The kalpas are in turn divided into a varying number of asamkhyeya (translated as incalculables or innumerables). The number of asamkhyeya in a kalpa is related to the career of the *bodhisattva* (aspiring Buddha, potential nirvana attainer),[7] i.e., his progress toward Buddhahood. In these cosmologies the cycles proceed endlessly, the only escape from the continual cyclicality being nirvana.

With respect to time itself, there is considerable variation among the various sects as to what aspects of time are real or exist, just as was the case with the six major Hindu schools. There seems to be general agreement that the present is real, which is consistent with the characteristic Buddhist sense of being in the here and now. Much of the controversy is devoted to determining what aspects of the past and future should be considered real along with the present. A discussion of the details of these various and extensive philosophic fine points is out of place here and is available in other texts.[8] Perhaps a flavor of the thinking involved can be captured by briefly considering some of the views of the great Buddhist leader Nagasena, which are expressed in his discourse with King Milinda. Nagasena states that there is past, present, and future time, and time which exists and time which does not exist. There are conditions in the past which have ceased to be; for them "time is not." However, there are also conditions which are producing their effect in the present or have the potential of doing so; for them "time is."[9] But the conditions are of this world and grounded in ignorance and thus subject to the samsaric chain. For those who can transcend the chain and reach nirvana, "time is not" (i.e., timelessness prevails).

Some interesting perceptions of time are also to be found in examining some of the Buddhist yogic practices. Much of this discipline was developed by the Yogacara school (Mahayana) which had some influence on the later assimilation of Tantrism discussed below. One fascinating meditative procedure is to start at a particular moment, then allow oneself to slip back to the next previous moment, then the next, etc., through past lives to the moment when life first burst into the world (i.e., at the beginning of a cycle), setting time in motion. One then reaches the paradoxical moment beyond which time did not exist, since nothing was manifested. The *yogin* (yoga practitioner) has thus effected an "emergence from time," in Eliade's words.[10] By so traveling back through time and reaching nontime, one reaches the unconditioned state and may thus enter nirvana.

At this point it is important to realize that the Buddha did not claim that the memory of former lives, etc., was the only means to reach nirvana. Indeed, according to him, it was perfectly possible to transcend time by taking advantage of the "favorable moment" and obtain instantaneous illumination. It is as if by daring to be fully in the living present moment the door to nirvana may open. Thus for many the ability to be totally in the here and now could be a stepping stone to nirvana. And the sense of timelessness that is experienced when one allows the living present to completely happen becomes permanent with nirvana.

Another interesting movement known as tantrism later began to work its way into the Mahayana Buddhist culture and existed side by side with earlier yogic forms, but it never actually replaced them. Typical of the tantric system is the concept of *sakti* wherein the female is regarded as the projected energy of the male. God and goddess are polar manifestations of a single transcendent principle; the male is identified with eternity, the female with time, and their embrace with the mystery of creation.[11] In tantrism the effort to transcend the cosmos was preceded by a long process of, in effect, cosmicizing the body and psychomental life, since it was from a "perfect" cosmos that one could then transcend the cosmic condition. Indeed the *Kalacakra-tantra* specifically relates the inhalation and exhalation in breathing first to day and night, then to fortnights, months, years, etc., to the longest cosmic cycles. Thus the yogin relives the great cosmic cycles and ultimately "emerges from time" and enters nirvana.[12]

The meaning of nirvana and how the indescribable is described may vary somewhat from school to school, but virtually all schools speak of reaching a state of unconditioned Reality. How it is described with respect to time and space is especially important here in terms of the fascinating view of the interrelation between the two. This perception of the interrelatedness of time and space is expressed very powerfully by D. T. Suzuki, 20th century scholar, in his discussion of the meditative state achieved in the practices of the Avatamsaka school of the Mahayana. There a condition is reached wherein there is no longer any consciousness of the distinction between mind and body or subject and object. In this situation:

> ...every object...is related to every other object and penetrated by it not only spatially but temporally. For this reason, every minute we live contains eternity. Eternal Now is our life; we do not have to seek eternity anywhere else but in ourselves. It is the same with the idea of space. The point I occupy is the center of the universe, and it is mine and with me that it subsists. As a fact of pure experience, however, there is no space without time, no time without space; they are also interpenetrating.[13]

This beautiful sense of the integration of time and space is also expressed by the great Tibetan Buddhist mystic Lama Govinda in his discussion of the experience of space in his meditative practice:

> And if we speak of the space-experience in meditation, we are dealing with an entirely different dimension.... In this space-experience the temporal sequence is converted into a simultaneous co-existence, the side-by-side existence of things into a state of mutual interpenetration, and this again does not remain static but becomes a living continuum in which time and space are integrated.[14]

Thus as a profound awareness of space is experienced in meditation, one arrives at an encompassing state of full presence where the whole stretch of time can be viewed, and time and space become one dynamic, interrelated whole.

The interesting similarity that these views have to the intimate connection between time and space suggested in relativity and quantum theory may be evident. A further discussion of these parallels will be given in Part III. But again, implied in the above statements is a special reverence for the living present, the eternal "now"; it is the framework of this profound momentariness that serves as the springboard to nirvana and its associated sense of timelessness.*

Summary

Mankind is deeply ensnared in the continual samsaric cycle of birth, death, and rebirth, endless in time. The escape from time is attained in nirvana. An essential feature of any path to nirvana is a deep sense of being in the vital present, which indeed has its own characteristic aspects of timelessness, which may be preliminary to the absolute sense of timelessness achieved in nirvana itself.

---

*This powerful sense of the living present is also characteristic of the Chan and Zen practices which were outgrowths of Buddhism in China and Japan, respectively.

An equally interesting view concerning time is the notion of the interpenetration of space and time as expressed in some of the Mahayana Buddhist thought and meditative practice. It may not be surprising that such a notion would emerge in Buddhism when one considers the interactiveness involved in the Buddha's concept of the twelve elements of Conditioned Coproduction discussed in this chapter. In any event there is a general similarity between the concept of intimate interrelation of time and space as expressed in the Mahayana meditative practice and that arrived at in relativity and quantum theories. The interpenetration of all objects also suggests a parallel with the overlapping of wave patterns in quantum theory. All of these time-related ideas will be developed further in Part III.

7

China and Taoism

Chinese religious and cultural thought are derived principally from three traditions: Confucian, Buddhist, and Taoist. For many scholars Confucianism is not actually regarded as a religion, but more as an ethical system governing behavior in all human interactions, from family and friends to national politics. Buddhism, which was discussed in Chapter 6, underwent some marked changes in the process of importation from India and Tibet and of assimilation into Chinese culture. While some of these changes will be discussed in this chapter, a more detailed discussion of Chinese Buddhism, though a rich tradition in its own right, is out of place here. Therefore, this chapter will first deal with some general observations about Chinese culture and history that will serve as a useful background and preface for some views of time as progressively synthesized from Confucian, Buddhist, Tao, and other traditions. This will be followed by an outline of some of the principal features of Taoist thought and its relation to concepts of time.

The Chinese Symbiosis

The two Eastern religions that have thus far been discussed originated in India. In treating Chinese religions it is important first to note some fundamental differences in the attitudes toward life and modes of thought between these two populations, both with such rich heritages.

While the Indians tended to engage extensively in philosophy, especially metaphysics, the Chinese were more practical, placing emphasis on moral behavior in familial and political relations and on the worldly ways of leading a successful life.[1] There was also a marked interest and respect for nature and the human body among the Chinese. The Indians had a predilection to subservience to some universal entity or concept, minimizing the relative importance of the individual, whereas the Chinese devoted their deepest reverence to their ancestors and to antiquity, namely the great figures and events in their historical past, as well as the sociopolitical hierarchy.[2] This made the Chinese excellent historians[3] in contrast to the Indians who were remarkably deficient in this respect. Both cultures on the whole were very tolerant religiously and exercised conciliation in accommodating differing viewpoints, but the Chinese generally went further, often with elaborate attempts to harmonize and transcend apparently conflicting concepts.[4]

It was such differences between Indian and Chinese cultural expression that were largely responsible for the acceptance of Buddhism in China for the most part in a modified form. The Buddhist concept of anatman (no-self) ran counter to the sense of self characteristic of the Taoist and Confucian traditions. Not only was the idea of rebirth new to China, but also such notions as cosmology, worlds, and kalpas conflicted with the deepseated Chinese belief in China as the world center. Although the Chinese translated most of the Buddhist scriptures into their own language, there were many changes, adaptations, distortions, interpolations, and embellishments in translation. For example, in some Chinese sects the concept of nirvana was identified not with a state of mind but with paradise or the "Pureland," which, as mentioned earlier, became the name of a predominantly Chinese Mahayana sect. Nevertheless Buddhism gradually developed into a strong religious and cultural force in China and took its place beside Confucianism and Taoism.

In the Confucian system the Chinese monarch was a son of Heaven, providing the human link between earth and Heaven. This was the zenith of a social and political hierarchy characterized by propriety in all human relationships and deference for superiors and ancestors. This attitude was particularly apparent in the profound respect given Confucian literature, especially as embodied in the five great Classics (Wu Ching). For the Chinese these works were endowed with an authority possessed by no other books and offered precedents par excellence for all aspects of living.

Perhaps the most interesting of the Classics is the *I Ching,* the *Classic of Changes,* which describes a cosmological structure that encompasses both man and nature in a single system. It can be regarded as a manual for interpreting events and inner states of mind. A unique feature of the *I Ching* is its system of sixty-five hexagrams, which if properly understood supposedly reveal profound meanings applicable to daily life. A typical hexagram consisting of full and broken lines is shown below. Often the numbers involved in these figures were time indicators, telling something of the quality of a given moment.

A quite significant aspect of the hexagrams is the way they represent the famous Chinese cosmic concept of yin-yang through the use of the solid (yang) and broken (yin) lines.* Yang is male, active, and identified with heaven; yin is female, passive, and identified with earth. The yin and yang principles thus enjoy universal application and are regarded as explaining all being and all change by their ceaseless interaction. Von Franz mentions one such application regarding space-time, in which time corresponds to yang and space to yin.[5] The yin-yang principle is often symbolized as shown in Figure 5.

Apparently the *I Ching* was included with the other classics in about the 2nd century B.C. during the Han dynasty, as result of Taoist influence. This is but one example of the gradual symbiosis, despite much conflict, of the Confucian, Taoist, and Buddhist traditions. Thus by the 1st century B.C. the yin-yang concept along with the above three traditions embodied the mainstream of Chinese thought. As a result, Chinese concepts of time can be regarded as both linear and cyclical,[6] although in the opinion of philosopher Joseph Needham the linear dominates.[7]

---

*This concept did not arise from Confucianism, Buddhism, or Taoism, but is essentially a product of Chinese culture itself.

Certainly, if only because the Chinese were such accomplished historians, they must have had a sense of linear time. This is apparent in their records of social relations and events, and is particularly evident in astronomical calculations, for example, by the monk-astronomer, I-Hsing, who concluded that 96,961,740 years had elapsed since the "Grand Origin."[8]

At the same time, curiously, there was a component of cyclicality in the Chinese view of political history, the successive dynasties exhibiting a periodicity in their rise and fall. There was a cyclical view also that arose from the Chinese perception of nature and the functions of the human body, and as will be seen in the next section, this cyclicality is strongly supported by Taoist concepts. It is almost as if this mixture of linear and cyclical concepts of time is another expression of the primal yin-yang principle, with yang representing linear time and yin cyclical time. Perhaps a justification for associating yin with cyclical time may be found in the following treatment of Taoism which shares at least one common fundamental characteristic with the yin principle, its passivity.

Figure 5: Yin-Yang Symbol

Taoism, The Way of Return

According to tradition the essentials of Taoist thought were given by Lao-tzu, whose historicity is uncertain but is thought by many to have been a Keeper of the Archives in the court of the Chou dynasty. Disgusted with the decadence of the dynasty, he left, resolving to pursue a more wholesome secluded life. He headed west and, on reaching Han-ku Pass, was asked by the

Keeper of the Pass, to write a book on his accumulated wisdom before retiring to seclusion. In response he produced the *Tao te Ching*. *Tao* means way; *te* means power or virtue; and *ching* means classic. Thus *Tao te Ching* translates as "The Classic of the Way and the Power."[9]

Whether Lao-tzu is a valid historical personage and writer of the *Tao te Ching* is in a sense relatively unimportant, for the significant thing about the book is its ideas. It upholds the virtue of anonymity, which may explain why the text cites no persons, dates, places, or events.[10] Such anonymity is consonant with the fundamental concept of enlightened inaction, passivity, and laissez faire that characterize the Tao philosophy. This is expressed in Lao-tzu's doctrine of *wu wei* (not doing), which is applied to everything from personal relations to politics and war.

One of the ideas useful in applying wu wei is to perceive that every action provokes a reaction, more often an undesired one. It is better to do nothing about evil and let it run its self-defeating course. For every time an attempt is made to correct evil, the evil only becomes reinforced by the importance that has been accorded it. In the words of the *Tao te Ching:* "The more laws you make the more thieves there will be."[11] Thus the best thing to do is to return good for evil, or as Lao-tzu puts it, "requite hatred with virtue (te)."[12] This sounds quite similar to the teachings of Christ.* But Taoists see this behavior as practical and the best way to get people to do what you want, not as a holy duty. Thus inaction succeeds by *being* rather than doing, by attitude rather than act.[13]

The concept of wu wei finds a much more universal expression than simply with respect to action and reaction per se. Taoism can be regarded as a viewpoint of complete relativism. For example, the idea of beauty generates the idea of ugliness, the idea of moral virtue produces that of wickedness.[14] Those who conform to morals resent those who do not, and making a morality of amorality is the other side of the same coin.[15]

Perhaps it is clear by now that what is intended is an enlightened inaction. Wu wei does not necessarily mean to avoid all ac-

---

*For some remarkable parallels between the New Testament and statements in the *Tao Te Ching* see H. Welch, *Taoism, the Parting of the Way* (Boston: Beacon Press, 1957), pp. 5-6.

tion but all aggressive and manipulative action. It means do nothing that is not natural and spontaneous. Mankind needs to recapture Eden with its indifference to good and evil.[16] A recovery of naturalness and free flowing native instincts underlies virtually all Taoist insights. The teachings of Lao-tzu were thus intuitive, with little credence being given to national structures, formal logic, and systematic philosophy. Indeed the best symbol for Tao is water:

> That which is best is similar to water; Water profits ten thousand things and does not oppose them. It is always at rest in humble places that people dislike. Thus it is close to Tao. [Tao te Ching, Chap 8][17]

The Tao is the mysterious quiet that pervades all nature. As will be evident, this is reflected in the Taoist cosmology, which reveals two aspects of Tao. There is an *apparent* aspect manifested by the order of the universe and an *absolute* aspect which is the Essence from which the order arises. This Essence or Absolute Tao is often referred to as the Nameless or Non-Being. This is nonbeing in the sense that it cannot be classed with anything we know as being; it is rather "no-thing":

> The Tao that can be Tao'd is not the Absolute Tao. The name that can be named is not the absolute name.[18]

Consistent with the pervasive sense of nature, the Nameless or Non-Being (Absolute Tao) is often symbolized by the "Mysterious Female" or by the "Valley Spirit." Thus a birth motif is evident wherein the Nameless or Non-Being gives birth to Being or the Named, which in turn produces Heaven and Earth and the ten thousand creatures, representing living creation. The Absolute Tao never dies; it is outside of time, but the process of creation arising from it and the subsequent decay and resolution are regarded as cyclical. Just as in creation all things emerge from the Tao, in the decay all things return to the Tao. This idea, along with a beautifully succinct summary of the Taoist belief, is given in the following quotation from Chapter 40 of the *Tao te Ching:*

> In the Tao the only motion is returning:
> The only quality, weakness.
> For though Heaven and Earth and the Ten Thousand

> Creatures were produced by Being,
> Being was produced by Non-Being.[19]

From the foregoing remarks it is apparent that one significant time-related property of the Tao doctrine is its cyclical view of the universe. Indeed according to Needham: "Nothing could be more striking than the appreciation of cyclical change, the cycle-mindedness, of the Taoists."[20] Scholar Holmes Welch similarly observes: "Tao's most important manifestation in the world is the cyclical pattern of change."[21] Although in the latter's opinion the cyclic pattern is governed not by time but by cause and effect,[22] time undoubtedly elapses in the course of the pattern. In contrast to the Hindu system, there is no specified or measured time duration for the cycles. Therefore while there is undoubtedly a cyclical view of time implicit in the Taoist concepts, it is at a much more nebulous, mysterious, and unspecific level than with the Hindus.

A second important time-related characteristic is the timelessness that can be associated with Non-Being, the Nameless. As mentioned earlier the Absolute Tao is outside of time; being nameless, it is also timeless. This property of timelessness bears some general similarity to that of the Hindu Brahman, who is indescribable and ineffable and thus also nameless. Indeed there are roughly the same gross cosmological characteristics in the two religions: a cyclical world pattern somehow sustained by a timeless, transcendent, and indescribable Reality from which the world emerges in creation and to which it returns in decay. However, in the case of Taoists an air of diffused natural beauty, living serenity, and immanent mystery pervades the cosmic picture.

Summary

We saw that by 100 B.C. a composite of the yin-yang principle, Buddhism, Confucianism, and Taoism was generally integrated into Chinese thought. Thus both linear and cyclical views of time were exhibited in Chinese culture. The linear arose largely from a deep commitment to social and political history, while the cyclical arose not only from history from the point of view of the periodicity of the dynasties but also from the rhythms of nature and the human body as well as Taoist influences.

In the Tao tradition there was a clear motif of cyclicality in the behavior of the world, created from the Absolute Tao and ultimately receding back to it. But the Absolute Tao is outside of time; it is the timeless Non-Being that sustains the universe. Therefore, both cyclical time and timelessness are apparent in the Tao system, but qualified by a nebulous sense of pervasive mystery not exhibited in the Hindu system.

8

From Cyclical Rituals to Judeo-Christian Linearity

Certainly primitive man and woman must have developed a profound awe and respect, if not love, for nature and its yearly regeneration which brought forth the fruits that sustained them. Although nature had its dangerous and threatening aspects, they came to understand these and either to avoid them or deal with them. By and large they made their peace with the power, beauty, and terror of nature and in so doing very closely identified themselves with its cycles: the changing seasons, the wax and wane of the moon, and the daily passage of the sun.

However, they also had to face a much more personally terrifying natural fact, that of their own ultimate death. As growing children, they no doubt noticed their elders dying and gradually were forced to conclude that that too would be their fate. It was thus quite understandable that they looked to nature with its yearly rebirth for an answer, for a consolation, for a security against the terror of death and time. Since nature is reborn with periodic frequency, perhaps they too could experience this rebirth by identifying themselves with its cycles closely enough. Thus in their most religious moments it is understandable that they might be strongly influenced by a cyclical concept of time.

On the other hand in their daily pursuits they were able to observe progressive (non-cyclical) changes, for example, as they constructed tools or a domicile or as they watched the continual growth of a tree through the seasons. Thus with some awareness of this sequential progress they may have had at least some sense of linear time in their prosaic day-to-day living. This is not to mention their awareness of the aging of themselves and their neighbors.

This contrast between these two time views (cyclical and linear) of archaic peoples is clearly expressed in the work of Mircea Eliade.[1] Basing his conclusions on common features found in many primitive religions throughout the world, he sees an interesting pattern of cyclical religious behavior among these traditions. While it cannot be claimed that all archaic religions conform to such a pattern, nevertheless among a widely distributed number of early cultures the case is made for a deep, regularly recurring need to return to some mythical beginning.

Eliade associates what he terms "sacred" time with such cyclically governed religious times and "profane" time with ordinary daily temporal existence. According to him, the periodic religious experience of rebirth for these archaic peoples was extremely profound. It was usually based on a ritualistic yearly repetition of some mythical creation act, often involving some hero-god who brought about creation and order by fighting and overcoming the forces of darkness, evil, and chaos. The power of the ritual lay in the profound belief of the participants that they were in reality reliving the creation act, or what Eliade terms the cosmogonic act.[2] These ritual acts, therefore, brought about a coincidence of the mythical instant with the present moment and thus were able to erase profane time, which was regarded as unreal.[3] Therefore, only the sacred time of reliving the cosmogonic act and its consequent experience of rebirth was real.

Often, but not always, this ritual imitation of the primordial act of some mythical hero, god, or archetype was strongly integrated with their instinctive reverence for the natural seasons, and was thus a yearly event. There was a deep and total sense of renewal because, as a result of the ritual, all past miseries, sins, and tragedies were erased. Any consciousness or consideration of the sequential accumulation of these burdens, or past events in general, was erased. Thus, using the term "history" for the awareness of such a linear sequence of events, Eliade maintains that the formula for tolerating the "terror of history,"[4] and the attendant terror of death, was a cyclical regeneration accomplished by imitating some divine mythical model of creation. This regeneration was complete; it was a *creation,* not simply a restoration.[5] Therefore, for these early peoples, time helped make possible the appearance and existence of things, but it had no ultimate effect on their existence because time itself was periodically regenerated.[6] For them the real world was one which periodically returned to its eternal anchor point at the archetypal primordial creation instant.

Often the experience of regeneration was extended to other significant occasions besides a yearly ritual of creation. That is, the creation myth served as a replication basis for other "creations," such as births, marriages, healings, the installation of a new ruler, etc. Thus among many archaic cultures every ritual had a divine model: as the gods did, so must humans do.[7]

The practice of reproducing the cosmogonic act of a mythical archetype had not only a temporal aspect but also a spatial aspect. The cyclically repeated rituals were always performed at the same hallowed spot. Regardless of what marked this location, be it a monolith, icon, or altar, the place was looked upon as the Center, the cosmic Center where the mythical creation act occurred. This notion was often extended to such acts as colonizing a new territory, wherein the act was regarded as transforming chaos to cosmos. The resulting houses, temples, or city were endowed with a reality of being at the Center, the "center of the world."[8]

One of the most fascinating features of this cyclical religious behavior was its worldwide distribution. Although the details of the practices may vary tremendously, at least significant features and components of the basic structure of such ritual are to be found with many primitive peoples throughout the world: the Polynesians, Australian Aborigines, North American Indians, archaic communities in the Middle East, Iran, and India. However, in the case of India this refers to early Vedic times because with the later maturation of Hinduism an escape was offered from this cyclicality, as was true of Buddhism also (see Chapters 6 and 7).

At this point it may be obvious that even in today's modern Western world there are significant remnants of such cyclical religious practices. It is, of course, seen in the celebration of Christmas, Easter, the Passover, and Ramadan. The Sabbath is an imitation of God who rested on the seventh day of creation.[9] A strong sense of renewal is quite apparent with the inauguration of a new American president. Actually to this day in parts of Iran and Iraq the creation act itself is repeated.[10] Traces of the sense of a cosmic Center may be found in the reverence attendant in the dedication ceremony of a significant building, bridge, or highway: the creative act is being repeated at this particular place, bringing "order" out of "chaos."

The Israelite Path to Linear Time

Despite their periodic indulgence in these remnants of cyclical religious time, modern Western peoples function primarily on a linear time basis. This is in large part because they generally derive from the Judeo-Christian religious tradition, a tradition which developed a linear history, and hence ultimately a sense of linear time. Although there were other ancient cultures with histories, e.g., the Egyptians, Iranians, Greeks, and Babylonians, it is the Hebrews and later the Christians who provided the Western world with its most clear-cut source from which a concept of linear time gradually grew.

As we will see, there were significant differences between the early Israelite religious history and other ancient religious traditions. A very important difference is that with the Israelites there was not just one divine event which was periodically celebrated, but several. These, along with many other momentous episodes in Israel's spiritual history, make up almost a 2000-year dialogue with God (Yahweh). The dialogue can be said to begin with Abraham, the first of the patriarchs, when Yahweh promised that he would be the father of many nations and that his descendants would possess the land of Canaan. In return Abraham rendered total faith and devotion to Yahweh.* This faith was later demonstrated with Abraham's willingness to sacrifice his son, Isaac.

It is out of place here to recount the many sacred events occurring with the other patriarchs, Isaac and Jacob, as well as with Joseph in Egypt. However, it is necessary to cite what is generally considered the most significant episode in Israel's interaction with Yahweh, the Exodus from Egypt, led by Moses, and the

---

*Such reciprocal commitments involving Yahweh were known as covenants. There were many such covenants and covenant renewals in Israel's development.

receipt of the Ten Commandments (often known as the Decalogue). The Exodus story stands out as a bedrock for the faith of the Israelites.

Later Joshua led the successful invasion of the Canaanite lands, and then led the Israelites in a renewal of the covenant with Yahweh at Shechem. Many events in which Yahweh is believed to have intervened for Israel's salvation also occurred under a series of subsequent leaders, known as judges, during the period when Israel was attempting to maintain and expand its hold on Canaan. Still other events took place under the kings and with the emergence of the prophets, after Israel became a monarchy in about 1000 B.C.

These brief comments about Israel's early history are intended to show that there were a number of sacred occasions in its long, almost continual dialogue with Yahweh, several of which became the subject of a yearly religious celebration. However, it is important to understand how these celebrations originated and developed.

After Moses led the Exodus from Egypt and Joshua directed the successful invasion of Canaan, the Israelites settled there and began the change from a nomadic to agrarian life. The people began to assimilate many of the agrarian customs of Canaan, and among them was the celebration of festivals at times of the year governed by the Canaanite calendar, based on the natural agricultural seasons. As in the case of the primitive communities discussed earlier, in these yearly religious festivals the original moment of the divine event being celebrated was relived with a total spiritual absorption. These occasions were characterized by a profound sense of joy and gratitude to Yahweh for his intervention to save his people.

However, there was one extremely important distinction between the Hebrews and the Canaanites. Though they may have adopted the Canaanite festivals and celebrated them at the appropriate seasonal times, the Israelites gave the celebrations an entirely different purpose and meaning. The festivals were not to give thanks to the Canaanite gods for some blessing of nature, but to celebrate a given sacred occasion in which one God, Yahweh, intervened in Israel's history for its salvation. They worshipped a God of history as opposed to one of nature. The agrarian Canaanite festivals were therefore "historicized" by the Israelites, who were not bound by nature but by divine historical events.[11] Well-known examples of such festivals are the Feast of the Unleavened Bread, associated with the Passover celebrating the Exodus, and the Feast of Tabernacles, celebrating the covenant renewal at Shechem. The celebration of such interactions with Yahweh accumulated to form a succession of events from which no single component could be omitted and from which later the Israelites gradually began to develop a sense of sequential linear history.[12]

Initially, the festivals were regarded as a series of separate events or "times," for the early Israelites had little comprehension of chronological time in the modern sense. In fact, the Hebrew language lacks a word for time in this sense, or for history. For them time generally meant "event," "moment of time," or "period of time."[13] As in the famous passage in Ecclesiastes (Eccles. 3.1 ff), there is a time "...to keep silent, to speak, to love, to hate," etc.

A particularly important biblical meaning for time, expressed in Greek by kairos,* had the general purport of "decisive moment."[14] A sense of fate and crisis was a strong component in its meaning. It was believed that it was God, Yahweh, who governed and dispensed these crucial moments. The hour of death is fixed by God. In this sense kairos naturally developed a judgmental aspect, and became an established term in many scriptures. Thus it could mean the "time of judgment" or "last time."[15] There were other important words for different aspects of time, such as chronos, which could mean time in general, or period of time, lifetime or age.

A further study of the various words for time in the Bible is impossible here,[16] but we can say that essentially the early Israelites had little idea of time or chronology in the modern sense. They thought in terms of a series of many times, and the very important ones were celebrated by their religious festivals.

However, starting in about the 10th century B.C. during the

---

*The use of Greek is especially important because of the *Septuagint,* which was a Greek translation of the Old Testament begun by Jewish scholars in Alexandria (Egypt) in the 3rd century B.C.

early period of the Israelite kings, the earliest writers of the Old Testament began to assemble the history of Israel's interaction with Yahweh. Even after the earliest of these scriptures were set down, it was still a gradual process by which the people began to identify their series of festivals and divine events with a linear history as we know it today. One reason it was slow in coming was that the priests and religious leaders wanted the events to maintain their spiritual vitality and did not want them to be regarded as something that had slipped into the irretrievable historical past.[17] In any case according to the Old Testament scholar Gerhard Von Rad, this development of linear history which Israel bequeathed to the West was one of its greatest achievements.[18]

Once the idea and perception of a linear history developed in the Hebrew culture, a foundation was laid for the growth of a linear sense of time. The linear sequence of sacred events in Israel's past served in a sense as a model and provided a basic apprehension of linearity, which was gradually projected into daily life, ultimately leading to a notion of linear time. It was a slow process, starting with what was a historical, theological series in the biblical era leading to a chronological time continuum centuries later.

There is one extremely important feature which underlies the Hebraic evolution into linear history, as well as the Christian continuance of this, soon to be discussed. It is relevant to the question: What made this evolution possible in terms of the Jews and Christians facing Eliade's terror of history? It will be recalled from what we discussed earlier that primitive peoples dealt with this terror by periodically endowing the primordial creation act with a living reality in which they totally immersed themselves, thereby erasing the burden of all past sins and tragedies. So in gradually giving up this cyclical antidote and launching into the unknown with linearity, what did the Hebrews, and later the Christians, find to replace it?

Eliade's quite sensible answer is *faith*.[19] The first virile seed for this was planted with Abraham's willingness to obey Yahweh's will and sacrifice his son, Isaac.[20] This faith found its firmest grounding in the Exodus experience. Looking back in the light of the Exodus, the Old Testament writers could see a growth and maintenance of faith through the sequence of patriarchs and beyond. Thus it was basically faith in Yahweh and the covenants made with him that guided the Israelites, and for them made possible such a spiritually powerful sequence of salvational events.

It was this same grounding in faith that made it possible for the Christians to face the terror of history and not resort to the repetitive renewal process of archaic peoples. With the Christians there were also many important sacred events, compressed, of course, into a much shorter period of time. By far the most important of these was the death and resurrection of Christ. In conjunction with the faith in this central divine event, as well as others in Christ's ministry, was a sense of linear history and time, inherited from the Jews and naturally maintained and propagated by the Christians.

In my opinion Eliade's association of faith with linear time is a profound insight. Why this may be so can perhaps be understood with some reflection. If we think about it, it may become apparent that in reality to live in linear irreversible time, and to live fully, requires an enormous amount of courage. There is an open-ended chance-taking and vulnerability demanded of men and women in living with linear unidirectional time. They are riding the arrow of time full tilt into the fog of the trackless future, not knowing with any certainty what will come next. This requires a visceral faith, conscious or unconscious, a faith for the Jews and Christians in a God for whom anything is possible. One who can deliver his people from the Egyptians and lead them to a homeland, as well as one who can move Christ to say:

> Have faith in God. For verily I say unto you, that whosoever shall say unto this mountain, Be thou removed, and be thou cast into the sea; and shall not doubt in his heart, but shall believe that those things which he saith shall come to pass; he shall have whatsoever he saith. [Mark 11:23]

In addition to what we have already discussed, there is another basis for the growth of the notion of linear history which arises from two basic ideas in the Old Testament tradition. The first is the belief in a creation as expressed in Genesis, but a creation which was not repeatable, and not ritualized or celebrated (except indirectly by the Sabbath). The second idea was the expectation of a Messiah (discussed below) which was a heritage from the period of the kings and prophets and began to be expressed most strongly during the exile of the Jews in Babylonia. These two ideas, one past, the other future, undoubtedly also impressed some sense of linear time onto the Jewish mind. Similarly the Christians believe in an unrepeatable creation and an end with Christ's second coming, which also engenders a sense of linearity. Let us now pass on to a more detailed discussion of the biblical past and future in terms of beginnings (Creation) and endings.

Beginnings and Eschatological Endings

Another important time-related aspect of the Judeo-Christian tradition is the notion that there was a divinely created beginning. Most citizens of the Western world are familiar with the first words of the Old Testament:

> In the beginning God created the heaven and the earth. And the earth was without form, and void; and darkness was upon the face of the deep. And the spirit of God moved upon the face of the waters. And God said, Let there be light and there was light. [Genesis 1:1-3]

There are three features about this quotation and the ensuing verses describing the creation of the earth, nature, and man which are of concern here. The first is the simple statement that there was indeed a beginning. Secondly, accompanying this beginning was light; indeed the production of light occupied the first day of creation. The third is the fact that the creation was done in stages, that is in six days, the last of which saw the creation of man.

While these three features are of interest for later comparison with corresponding facts of physical cosmology, the origin of the Genesis creation story must be examined more closely. Most biblical scholars believe that some of the themes found in the story were borrowed from earlier Mesopotamian and Canaanite myths, but appropriately altered to fit the theological needs of the Old Testament.

Among the most likely candidates for an earlier source is the Mesopotamian creation epic, known as the Enuma Elish, which begins with chaos represented by the sea with no land at all. Full details of the epic are out of place here. It is sufficient to mention that ultimately chaos is personified by the female deity, Tiamat, in the form of a dragon. In mortal combat, Marduk, the creator deity, slays her and then creates the visible world. The earth is a disk resting on the chaotic abyss of the ocean.[21] A similar pattern is apparent in the Canaanite myth described in the tablets of Ugarit where the creator deity is Aleyan Baal.*

The writers of Genesis used motifs from these myths to provide a setting and starting point for the Old Testament. In a sense the Genesis account is a criticism of the Mesopotamian and Canaanite myths because in this account there are no male and female divinities and no deification of such things as chaos, water, light, or animals.[22] For the biblical writers there was only one deity, God, who created all.

Thus, the comparatively unique characteristic of the Genesis account is its unqualified monotheistic nature: the beginning was brought about by the creative work of the one God. This beginning initiated a course of linear history for both the Jews and the Christians. In this history, as discussed earlier, God intervened—in a series of sacred events in the case of the Hebrews, and for the Christians with the birth, ministry, death, and resurrection of Christ. An interesting aspect of this linear record can be found in a comparison of the Jewish and Christian calendars, which reflect their respective views of the sweep of theological and redemptive history. While the Hebrew calendar begins with God's creation of the world, the Christian calendar begins with the birth of Christ, and time is measured in each direction (A.D. and B.C.) from this "center."

According to biblical scholar Otto Cullman, the early Christians as well as the then contemporary Jews felt that time was an endless flow.[23] They saw redemptive history divided into three ages: before creation, from creation to Judgment Day, and after Judgment Day. What is termed Judgment Day here has been

---
*Ugarit is an ancient city on the Syrian coast, at the modern-day Ras Shamra site.

called many things: the *Parousia* (meaning generally divine appearance or presence), the Coming of the Kingdom of God, the Day of Redemption, etc. For the Jews it was the coming of the Messiah, the authentic descendant of David, because they believed that Jesus did not fill this role. For the Christians it was the second coming of Christ. Cullman suggests that superimposed on this threefold division of the ages as indicated in Fig. 6 is a twofold division brought about by the central redemptive event. This again is the "Christ event" for the Christians, but for the Jews it is yet to happen with the coming of the Messianic Age.

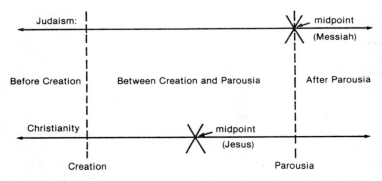

Figure 6. From Oscar Cullman, Christ and Time: The Primitive Christian Conception of Time and History *(Revised Edition), translated from the German by Floyd V. Filson. Copyright (c) MCMLXIV W.L. Jenkins. By permission of The Westminster Press.*

Although many scholars disagree with Cullman's viewpoint and schematization of biblical time scales, it has provided a perceptual position against which other theories of biblical time have been contrasted. It has often served as a starting point for the arguments of many writers in biblical study. Perhaps the most comprehensive criticism of Cullman's viewpoint comes from Marsh,[24] who holds that the concept of time as an endless flow cannot be justifiably extracted from the Bible. Just because God's time, eternity, is different from ours does not necessarily mean it is endless time. Marsh also points out that abstract, chronological time is a modern concept, not a biblical one, so that projection of such a modern view on to biblical history in terms of the schematization in Fig. 6 is questionable. Lastly he maintains that there is ample biblical evidence to support the contention that the Christ event was not a midpoint but an end point. That is, all of God's spiritual work was completed in Christ; all we have to do is "catch up" to that fact.

The foregoing discussion of the controversy concerning the attitudes of the early Christians towards time, creation, and the end gives us a glimpse of some of the variations in biblical interpretation on these subjects. Nevertheless there is a general agreement that cyclical time was refuted and some form of linear time accepted, though probably not in a modern sense. Also there is a clear belief in a creation and a Day of Judgment, but as to whether there was timelessness antecedent to creation, for example, or an endless flow of time preceding creation is not settled. As we will see in the next chapter, St. Augustine came down clearly on the side of timelessness.

Undoubtedly much of such controversy arises from different interpretations of the meaning of the resurrection and events in Christ's ministry, as well as from the import of his second coming in terms of what is known as eschatology. This is the area of theology concerned with the consummation of the divine purpose in history, the final judgment, and disposition of individual souls at death. In Christian eschatology there is considerable variance of opinion among scholars as to the time of arrival of the Kingdom of God. There are those who hold to an older traditional view that this is to occur in the future. Other scholars such as C.H. Dodd, who express what is often termed a "realized eschatology," maintain that with Christ God aleady introduced his Kingdom.[25] From Christ's time onward it has been here, ready for those who would open their hearts to it.

Still other scholars hold that either time should not figure in Christian eschatology at all,[26] or that there is a curious mixture of the present and the future in the Christian eschatology. With the appearance of Christ a decisive battle in redemptive history was won. However, the entire war against evil was not yet won, even though the decisive battle had fixed the ultimate outcome.[27] That outcome, the winning of the war with evil, would occur with Christ's second coming. Thus there is a kind of anticipative yet partially realized eschatology (often termed "proleptic eschatology"), which makes for a strange blend of future and present.

There are predictions of the scenarios for the two redemptive events mentioned above: the coming of the Messianic Age for the Jews and Christ's second coming for the Christians. These are to be found in apocalyptic scriptures of the Bible as well as other such scriptures that were not endorsed by the Church as canonical. Apocalyptic literature, almost invariably written under a pseudonym, is characterized by much symbolic imagery and usually foretells an imminent cosmic cataclysm wherein God destroys the evil ruling powers and raises the righteous to everlasting life in his Kingdom.

Daniel in the Old Testament and Revelation in the New Testament* are the two books of the Bible which are well-known examples of such apocalyptic literature. Although there are other apocalyptic passages both in the Old and New Testaments, e.g., Isaiah 24-27 and Mark 13, Daniel and Revelation essentially represent a mature embodiment of this type of scripture. The purpose of such writing was to give courage to the faithful in the face of suppression and persecution. In the case of Daniel it was designed to provide solace and inspiration for the Jews whose worship and law were being interdicted by Antiochus Epiphanes IV, Seleucid ruler of the Syrian empire, a segment of the earlier empire of Alexander the Great.[28] It was at this time in the second century B.C. that Judas Maccabaeus led the revolt against Antiochus's attempts to force Greek deities and culture on the Jews. Revelation was written, according to biblical scholars,[29] to reinforce the faith of Christians in the first century A.D. who were facing the threat of persecution if they did not yield to the practice of Roman emperor worship.

With these theological motivations for writing the two books in mind, we can look at how they envision the end of the world. As mentioned earlier the evil rulers and their empires will suffer God's Judgment and be destroyed. Then in Daniel 12:2-4 it is

---

*Revelation is an exception to the general rule that most apocalyptic literature is pseudonymous. It was written by an inspired Jewish Christian named John, who most scholars believe was not John the Apostle, but another John, perhaps one known as John the Elder.

described how the dead shall rise and be judged, with the good receiving everlasting life, and the evil, everlasting shame and contempt. This incidentally is one of the earliest expressions of the belief in resurrection of both righteous and wicked. As to the date this is to happen, verse 4 tells us that Daniel is to "shut up the words, and seal the book even to the time of the end"; thus no time is given.

In Revelation, after God's wrath visits the world with a series of plagues coming in groups of seven, chapter 17 describes the judgment of Babylon, which represents Rome and is personified by a harlot who has seduced mankind with her cult of emperor worship. Then finally (Rev. 19:20) "the beast"(representing the Roman Caesars) and the "false prophet" (the priests of emperor worship) are "cast alive into a lake of fire burning with brimstone." Later (Rev. 20:20) the devil himself suffers the same fate, after which all the dead arise. Those who are not "written in the book of life" are also thrown into the lake of fire (Rev. 20:15), suffering a second death. The others are blessed with eternal life in the "New Jerusalem," representing the Kingdom of God and fulfillment of his purpose.

The apocalypic scenarios in Daniel and Revelation have in common the forecast of a divinely imposed End preceded by cataclysmic natural events. The End is profoundly judgmental, redemptive, and soteriological (salvational) in nature. However, of specific time-related interest here is the simple fact that an end is predicted. It is obvious that the characteristics of time, if indeed there is time, "after" the End is quite unclear in these biblical accounts primarily written for theological purposes. However, further discussion of the relation of the biblical beginnings and endings to their physical counterparts must wait to be taken up in Part III.

Summary

It was shown how many archaic peoples were considered to have lived cyclically in their religious life in accordance with the natural seasons, believing the only real or sacred time was time when some cosmogonic and mythical creation event was re-enacted. Such a cyclical nature-based religious life was practiced by the Canaanites when they were invaded by the Israelites, who adapted such yearly celebrations to their own use. That is, they used these events to praise a God of history, not of nature, one God who on many occasions intervened in human history to guide his people. From this chain of such divine occasions a sense of linear history and time gradually developed. A sense of linearity was also derived from the Judeo-Christian beliefs in an unrepeatable creation and some form of a redemptive end yet to come.

We saw that for both the Judaic and Christian traditions the beginning of creation is described in Genesis. This account tells us that there was indeed a beginning, that it was accompanied by the production of light, and that it was accomplished in stages. Finally in the Judaic and Christian traditions the apocalypses of Daniel and Revelation, respectively, give us a prediction and description of the end which, however, is primarily judgmental and soteriological in nature without clearly specifying when the end will come or whether it will also bring the end of time.

9

Four Western Theologians

We continue the examination of time-related Western religious views with selected distillations from the thought of four Western theologians: St. Augustine, Alfred North Whitehead, Martin Buber, and Teilhard de Chardin. It will be evident that each of these men expresses a unique view of time or time-related perception. Together they embody a rich spectrum of religious insight bearing on time that will prove extremely useful in drawing together the conceptual threads thus far discussed into some kind of a harmonious fabric.

To some it may seem strange that Whitehead is included here since he is probably most often regarded as a philosopher. However, he studied theology over a period of some eight years, and while he ultimately decided not to join any church, he was possessed of considerable spiritual insight. His metaphysics is an impressive synthesis of religious as well as scientific components and is among the strongest bases for a whole area of religious thought known as process theology. Accordingly, it is judged quite necessary to include Whitehead's singular ideas with those of the other thinkers.

Actually a similar objection could be raised with St. Augustine, since much of his work with respect to time per se is of a philosophic nature. Nevertheless, his views of time, while at times somewhat philosophic, are firmly grounded in a profound faith and are strongly affected by this viewpoint.

Some of the thought of Martin Buber is included because of his deep understanding, his sense, and his expression of the living present. This understanding, along with a knowledge of his concept of "I-Thou" and "I-It," will prove most useful in our further discussion of physical and religious time comparisons in Part III.

No comment is necessary in the case of Teilhard de Chardin because the applicability of his views to religious time concepts will be quite apparent. With this introduction in mind let us proceed with surveys of the time-related thought of the four men.

The Questions of St. Augustine

This great defender of the Church was one of the most completely developed men in its history. At once an intellectual giant and a profound mystic, bringing himself from moral degradation to spiritual heights, he was a powerful mainstay of the Church during the perilous times of the progressive disintegration of the Roman Empire. Of his two greatest classics, *Confessions* and *The City of God,* we will limit ourselves primarily to a discussion of certain selected portions of Chapter XI of the *Confessions.*[1]

Although it is in this chapter that almost all of St. Augustine's comments about time are contained, he does make occasional allusions to time throughout earlier portions of the book. Here he expresses God's eternality as God's everlasting "today" that survives all of mankind's yesterdays and tomorrows as a continual living presence. He also emphasizes God's governorship over time and his eternity which has no past or future because it *is*.

Augustine's principal examination of time (Chapter XI) comes in the form of a one-sided dialogue with God, in which he frequently asks God questions and on occasion sets down answers, resulting either from his own logic or from divine inspiration. He starts by asserting that God's Word was expressed via the motion of something he created: motion subjected to the laws of time. However, God's Word itself is eternal and independent of its expressions found in creation. He criticizes those who idly speculate about what God was doing before creation, who wonder why if he is eternal did he not create something eternal. To them he answers: "How can anyone ask what you [God] were doing 'then'? If there was no time, there was no 'then,' " and suggests they be still, and listen if they would understand God's will and his creation. Although God was before time, it is not in time that he precedes it, for he made time. That is, he precedes creation and time, but not in a temporal sense. Thus before God erected the world and with it time, there was no time.*

Augustine then examines the three general divisions of time: past, present, and future. He goes to considerable length to demonstrate how only the present exists, and that only for an instant. Since only the present *is,* the past and future do not exist:

> How can...the past and future be when the past no longer is and the future is not yet. As for the present, if it were always present and never moved on to become the past, it would not be time but eternity.[2]

Finally, he pursues his queries to God in an examination of the measurement of time and finds that it must be measured while it is in process of passing, not before or after, for then it no longer exists. The question of measurement is presented in a way that somehow expresses its profundity:

> What am I measuring when I say either, by a rough computation, that one period of time is longer than another, or, with more precision, that it is twice as long?...While it was transient, it was gaining some extent in time by which it could be measured, but not in present time, for the present has no extent....We cannot measure it if it is not yet in being, or if it is no longer in being.[3]

In attempts to respond to these paradoxes, Augustine first speculates that time is some kind of an extension, perhaps of the mind itself. This leads him to examine the three mental functions, memory, attention, and expectations, as they relate to past, present, and future, respectively. He asserts that he really measures time in his mind. What transpired in the measurement makes an impression on the mind, which remains after the time of measurement ceases to be. He then concludes that it is the impression that he measures and not the thing itself: "Either, then, this is what time is, or else I do not measure time at all."[4] In other words, when Augustine measures the time between two

---

*The time-impregnated word "before" is used here for lack of a better word.

events, the present moment that existed at the first event is a recorded memory when the second event occurs, which in its turn becomes a recorded event. The time measurement is the difference of these two memory records of time.

Although Augustine is apparently dealing with some rather analytic aspects of time and its measurement, it must again be remembered that his entire discourse took place in the context of a dialogue with God. He asked God questions about these perplexing problems with a totally reverential attitude and sought answers with all of the spiritual insight at his command. His prime motivation was theological in that he wished to examine and contrast time in the inconstant created world of mankind with that in God's eternity. The feature of this contrast of most relevant interest to us in this book is Augustine's firm conviction that there was no time antecedent to creation. This is a religious view which presents us with an engaging comparison with the physical view of John Wheeler, both of which will be discussed further in Part III.

Alfred North Whitehead

After having enjoyed a very successful and productive career as a mathematician and educationist, Whitehead became one of the foremost philosophers of the 20th century. When he turned to philosophy and came to the United States, he developed one of the most comprehensive formulations of metaphysics to be offered in this century, encompassing a spectrum of religious and scientific insight. Thus his ideas on metaphysics continue to engage the interest of many thinkers, both philosophic and religious.

A persistent motif that characterizes much of Whitehead's thought is the intimacy and immediacy with which sense and perception interact with nature in a vibrant, advancing process. He balanced this by the notion that the natural world is guided toward a value-endowed aim and realization by a generally transcendent, as well as divine, influence.

The central notion of Whitehead's metaphysics is that the world can be described as a process of concrete, finite entities which are unique, complex, and interrelated.[5] These "quanta or atoms of experience" are called "actual occasions" and are the final real things of which the world is made.[6] The becoming of an actual occasion occupies and atomizes a particular region of the extensive space-time continuum, and each occasion mirrors a world of occasions from its perspective.[7] Thus an actual occasion requires a finite temporal interval for its realization. In this manner everything in nature experiences—from man to rocks—so that all of reality is interrelated and processive in character.[8]

A fundamental and complementary concept in Whitehead's scheme is that of "eternal objects," which are pure ideas, ideal potentialities, or possibilities for realization.* There are all varieties of "eternal objects": shapes, colors, numbers, sounds, feelings, etc. Eternal objects are the abstract pure idealic elements in terms of which actual occasions develop their identity and realize their novelty.[9]

A third important notion of Whitehead's system is that of "prehension," which connotes grasping, feeling, apprehending, or appropriating. Thus an actual occasion prehends or acquisitionally "feels" a certain pattern of eternal objects in its process of maturation. In brief an actual occasion is a prehensive, processive unification of a given pattern of eternal objects that is unique and peculiar to that occasion;[10] and this occasion is atomized in a portion of the extensive continuum.

The process of the maturation or "concrescence" of an actual occasion involves both the inclusion and exclusion of data from previous occasions, as well as from external objects. After an actual occasion concresces, it then perishes. For Whitehead the world is a process of occasions which arise and perish.[11] However, in a very real sense the actual occasion is not obliterated but is immortal. This is because it serves as a datum for subsequent occasions, which are part of a whole interrelated complex or "nexus" of actual occasions in the space-time continuum. A human being is a nexus or society of actual occasions as well as a series of actual occasions that make up personal experience.

---

*The concept of eternal objects suggests Plato's teaching that reality does not consist of tangible observable objects like houses and dogs, but the idea or "universal prototype" of the object. That is, the idea of a house exists independently of whether a particular house exists.

A final very important aspect of actual occasions is that each is somehow regulated by a "subjective aim." That is, the concrescence of an actual occasion is directed by this subjective aim, which governs the ordering of prehended eternal objects and the goal of the occasion, and is the factor by which the occasion is defined and maintains uniqueness throughout the concrescence process.[12]

The foregoing concepts dealing with Whitehead's metaphysical view of the world are necessary to understanding his ideas about religion and time. As a further preface to comprehending these ideas, it is useful to present some general observations of his views concerning the nature of God. Whitehead does not see God as omnipotent, because otherwise he would be responsible for the evil as well as the good in the world. Therefore: "The limitation of God is His goodness."[13]

God is complete only in the sense that the totality of his ideal vision "determines every possibility of value." It is as the source and provider of the ideal direction that God manifests his strength: "The power by which God sustains the world is the power of Himself as the ideal."[14]

God does not predetermine events. He only can provide an antecedent grounding which serves as a guide with some limitations qualifying every creative act.[15] It is in this capacity that God is interactive with creation. To Whitehead this interactiveness reveals that God needs the world as much as it needs him.

More specifically, in terms of the actual occasions, eternal objects, prehensions, and subjective aims defined above, God, using "the power of Himself as an ideal," furnishes a grounding and initial direction for the subjective aim of an actual occasion. He also provides the ideal conceptual ordering of the eternal objects prehended by actual occasions.[16] Thus he lures, not determines, the world of actual occasions towards the ideals and values ultimately possible for it.[17] It is through this mediative interaction and the resultant realization of God by actual occasions that God himself can be said to be processive. God uses and needs actual occasions "as an intermediate step towards the fulfillment of His own being."[18]

So God is not a despotic agency, he only persuades and lures.[19] God cannot prevent the occurrence of evil. Each actual occasion in its concrescent process has some freedom of self-determination. Therefore there is evil in the world because the occasions do not wholly conform to the ideal norms God conceives for them.[20] God's aim for every occasion is then the maximum fulfillment that can be attained in that given situation.[21] It is because of this falling short that God suffers, since he envisages the ideal.[22]

Having established that Whitehead does indeed express some unique and incisive religious thought, let us pass to some of his views concerning time. His perception of time and space is strongly colored by his visualizing nature as an ever-changing, interrelated organic unity. Since it is ever-changing, a thing is itself only just when and where it is; at other times and other places it is a different thing.[23] On the other hand everything is related to everything else so that:

> In a certain sense, everything is everywhere at all times. For every location involves an aspect of itself in every other location. Thus every spatio-temporal standpoint mirrors the world.[24]

Time is regarded by Whitehead as an intrinsic expression of reality, which to him is creative advance in nature. He therefore is convinced that time is in nature and not nature in time.* Time being in nature is consistent with his concept of "actual occasions," wherein he holds that events are atomic and need a finite temporal interval in order to display their pattern and achieve completion.[25] Whitehead does not believe in instants of time (i.e., points in time) or points in space. For him temporal advance is impossible in nature via a succession of instants, because nature could not advance beyond the immobile present instant on which it is "impaled."[26] Whitehead's complaint is that the mathematical concept of a continuum of points is given a role beyond its theoretical one; it is asserted to be a fascimile of the real spatiotemporal continuum in nature where no points are actually perceived.[27]

It is not surprising then that he also holds that actual occa-

---

*Remember nature being in time was the old Newtonian view, i.e., time went on its independent way, unaffected by and outside of nature.

sions require finite space and time to come to fruition and that time and space are interrelated: "There can be no time apart from space; and no space apart from time; and no space and no time apart from the passage of events in nature." (A. N. Whitehead, *Concept of Nature,* London: Cambridge University Press, 1920, p. 142.) It is interesting to note the broad similarity of this concept to the view of the early Israelites, in whose world events were the prime reality.

Because of this intimate interdependence of time and space in terms of extensive quanta of actual occasions in the four-dimensional continuum of nature, there is no absolute or unique physical meaning to the idea of simultaneity.[28] Even more interesting is Whitehead's imaginative speculation that the universe has evolved through the ages to its present four-dimensional space-time system. However, with the continuation of creative evolution this system may dissolve into other orders of space-time with other numbers of dimensions. Future beings may look back with contempt from their fifteen-dimensional viewpoint at our myopic four.[29]

Whitehead's concept of actual occasions in space-time as atomic quanta of experience alters somewhat the view of the reality of the past and future with respect to the present. Because of the extensive nature of time, the future is in some sense suggested by the temporal nature of the present, as the present was by the past. Recall that one of the characteristics of an actual occasion is that upon full maturation and subsequent perishing it is nevertheless "immortalized" as a component of a succeeding actual occasion. Therefore, for Whitehead the often accepted notion of the non-reality of the past and future must be qualified.

On the other hand, remember that each actual occasion is unique. This being the case it can never be repeated. Since as stated earlier no two events are ever exactly the same, and once having occurred are irrevocable, Whitehead thus shows that in his system the advance of time is irreversible.[30]

Martin Buber

In his studies as a young man, Buber proved to be an impressive interdisciplinarian. He did not believe in extensively restricted specialization, feeling that there was always some significant relationship between things and ideas to which men and women devote their attention and study. For him a broad approach led to an inner sense of unity and self-integration. Generally disdaining the practices of classification, systematization, and rigorous philosophical or rational thought, he might best be termed a phenomenologist or existentialist. However, in keeping with his cultural holism he did not wish to be labeled. Although many have called him a theologian, philosopher, or educator, all with very good reason, he personally felt his interdisciplinary efforts could best be described as "philosophical anthropology."

Buber's fascination with the Eastern religions, Hinduism, Buddhism, and Taoism, undoubtedly helped contribute to the exquisite, incisive, and poetic perception that brings such vitality and presence to his religious writings. These included a broad spectrum of works on Judaism, Zionism, and biblical interpretation. However, perhaps the most important source for his thought arose from his life-long efforts to interpret 18th century Hasidism in Eastern Europe.[31] For Buber, Hasidism is the greatest phenomenon in the history of the human spirit, being a society that lived so completely by faith. He has been criticized for not giving more studious attention to the orthodox Judaic aspects of Hasidism, but his profound sense for Hasidic spirituality constitutes an invaluable contribution to modern religious literature.

For the Hasids, according to Buber, holiness and spirituality entered all aspects of life. Life must not be compartmentalized into the holy and profane; it must be lived in the "here and now" in the spirit of faith which sanctifies each act with full presence and attention. It is this essential challenge of taking a chance on one's true feelings and stepping totally, without reservation into the living present that Buber addresses in his extensive and well-known writings on interhuman relation. Undoubtedly the most famous of these is *I and Thou*.[32]

In this work he sets down the essential elements necessary for genuine dialogue, I-Thou dialogue. There must be sufficient "distance" for the persons so engaged for them to have "space" to freely express their full individuality and personality. From this mutual posture each of the two parties turns to the other in full openness and readiness for completely spontaneous and unanticipated dialogue. Such a dialogue is a unique happening in that moment and will never happen in the same way again. All concerns about making an impression, looking good, or image building have evaporated; the two in dialogue are sharing themselves in the unplanned grace of the full living present. In Buber's words: "The Thou meets me through grace—it is not found by seeking.... The primary word 'I-Thou' can be spoken only with the whole being.... All real living is meeting."[33]

The Thou, the other, need not necessarily be another person; it could be dog, a tree, a work of art, or God. But in addressing Thou with one's full being one becomes I: "I become through my relation to the Thou; as I become I, I say Thou."[34] That is, in a dialogue the more totally I present myself, the better able am I to utter "Thou" with full grace and commitment.

Buber distinguishes two fundamental forms of relation, the I-Thou and the "I-It." In contrast to the "subject-subject" relation of I-Thou, the I-It is a subject-object relation. Here the subject uses, dominates (or tries to), or manipulates the other as if it were not a subject but an object. According to Buber "the primary word I-Thou is spoken out of natural combination, and that of the primary word I-It out of natural separation."[35] Thus the I-Thou word is exchanged in a totally natural relational happening, but the I-It occurs in a natural withdrawal or separation.

It should not be assumed from the foregoing that the I-It relation is necessarily evil. All people indulge in it in their conversations to a greater or lesser extent most of their lives. Most people can number on the fingers of one hand the times they have experienced an I-Thou dialogue. Buber clearly states that I-It is necessary in life, but he warns: "In all seriousness of truth, hear this: without 'It' man cannot live. But he who lives with 'It' alone is not a man."[36]

Before showing how some of the above remarks can be meaningfully related to time concepts, some of Buber's more philosophic views of time should be mentioned. He distinguishes between cosmological time and anthropological time.[37] The former is abstract and used as though the full sweep of time could be viewed by us in a relative way, even though the future is not yet available. The latter is "time in respect of actual, consciously willing man,"[38] and essentially has reality only in the past, since man cannot entirely will the future. The reason for claiming that this time has a real basis in the past is related to St. Augustine's thought concerning time measurement as grounded in the memory. As Buber put it: "As soon as we become conscious of the dimension of time as such, the memory is already in play."[39]

In this connection he further helps distinguish I-Thou and I-It in terms of the present and the past: "The 'I' of the primary word I-It...has no present, only the past."[40] On the other hand I-Thou exists totally in the present: "True beings are lived in the present, the life of objects is in the past."[41] A very important characteristic of the I-Thou relation is the fact that "the pure present knows no specific consciousness of time."[42] That is, as anyone knows, who has had an I-Thou experience or has for a period at least lived in the present, there is little or no consciousness of time. One is graced with the timeless eye of the living present. Certainly it is the ability to be wholly in the state of the timeless present that is a necessary precursor to many mystical experiences. Our most intimate glimpse of timelessness and eternity are afforded us through being totally in the "Now."

Another very important aspect of the I-Thou relation, which was mentioned earlier, is its complete uniqueness. Once it happens such a dialogue will never happen exactly the same way again. The subtle nuances, the aura, and singular flavor of the meeting can never be reproduced. This constitutes a very clear statement at the spiritually human level of the irreversibility of time, and bears some general similarity to Whitehead's position that actual occasions are unique and can never be repeated. This aspect of time's unidirectionality will be further treated in Part III.

Pierre Teilhard de Chardin

In common with the previous three men discussed in this chapter, this famous Jesuit thinker exhibited his versatility in several fields. These included paleontology and geology. Refusing to be a chaplain in World War I, he distinguished himself for valor as a stretcher-bearer, winning the French Legion of Honor. It was later in China that his interests in paleontology and geology flourished.

Surely it was his experience in these disciplines that contributed to the formulation of his unique perception of mankind's future evolution. He saw human evolution as an uncompleted process wherein man is mentally, socially, and spiritually evolving toward an ultimate spiritual unity.

The evolutionary process is essentially an internal one, and it does not proceed by God's external interference or manipulation. God fashioned the universe so that it evolves to its ultimate goal as a result of the inherent tendencies of matter. However, Teilhard still allowed for the occasional divine injection in the form of miracles, which he regarded as simply an enrichment of the general natural process.[43] In any event he views evolution as a progressive "complexification," starting with the electron and proton comprising the hydrogen atom, then more complicated atoms, then molecules (i.e., complexes of atoms), then biological organism, thence to man.[44]

The universe is regarded as a vast spatiotemporal system that by its nature is both organic and atomic. Everything in the world appears and exists as a function of this whole system. Teilhard asks the question: Is the operation and movement of this system aimless or directed? In particular, does the universe "show signs of containing within itself a favored axis of evolution?"[45]

He proposes to demonstrate that such an axis does exist, and he bases his exposition on "the principle that the greater coherence is an infallible sign of the greater truth."[46] Specifically, he attempts to show that such an axis may be defined in terms of three (perhaps four) successive theorems or postulates, each of which finds its basis in the preceding one. The postulates delineate a progression along an axis pointing toward a deeper fulfillment and realization for men and the universe, and, as stated by Teilhard, they are:

a) Life is not an accident in the Material Universe, but the essence of the phenomenon.
b) Reflection [his term for Man] is not an incident in the biological world, but a higher form of Life.
c) In the human world the social phenomenon is not a super-

ficial arrangement, but denotes an essential advance of Man.
d) To which may be added, from the Christian point of view: [Christianity and the Christian community] is not an accessory or divergent shoot in the human social organism, but constitutes the axis itself of socialization.[47]

Regarding the first theorem Teilhard claims that the reason that life is so rare in the universe is because, representing a higher form of cosmic evolution, it can come into existence only in privileged circumstances of time and space. Thus animate beings are not a fortuitous byproduct in the universe but the characteristic and specific higher aim of the universal phenomenon of evolution. He further maintains that the degree of consciousness attained by living creatures may be used as a parameter to estimate the speed of evolution.[48]

Therefore for Teilhard, life is the spearhead of evolution and can also provide a pointer to the direction of its advance. The pointer pointed for eons until a single ray of life broke through to man, passing through the "critical barrier separating the Unreflective from the Reflective",[49] the Reflective, of course, means thinking, self-aware humans or beings. This ray was only one among many previous attempts, but once having pierced the barrier, the dikes were broken and thought flooded the entire surface of the biosphere.

With this breakthrough the gradual process of collective Reflection (the society of all Mankind) began, analyzing, distilling, and resynthesizing knowledge and experience, but with a direction. Teilhard senses a directive flow toward a critical point of maturation corresponding to the concentration of collective Reflection at a single center embracing all the individual units (persons) on earth. Teilhard has often called this the "Omega Point."

Beyond this neither he or we can see, but obviously he senses further axial progress to another breakthrough. This time it would be a breakthrough of the centered and matured collective Reflections to a union with God or Christ, that is, to the Parousia (Christ's second coming) promised in Revelation, thus following a Christian-oriented axis.

Such a unique perception of the universe's evolution as a kind of ultimately spiritualized process has understandably aroused considerable discussion in the communities of religious and philosophic scholarship. Nevertheless Teilhard's thought exhibits obvious intuitive and spiritual insight and could provide a richer context for Christianity than has past Western thought since the Middle Ages.[50]

The obvious time-related feature of Teilhard's evolutionary picture is the linear unidirectional progress of time. In addition he sees this irreversible time as proceeding to an ultimate, essentially inescapable end. Whether or not time will end then is not specified—whether the Omega Point is Christian or not is not known—but an end in some form is a fundamental part of Teilhard's convictions. Finally, it is of interest to note that the notion of a purposive goal for evolution bears some interesting similarities to Whitehead's idea of God luring actual occasions to their richest fulfillment.

Summary

St. Augustine presented a very clear thesis, both religiously and philosophically based, for the reality of the present as opposed to the past and future. It was suggested that he was influenced in this view by his strong spiritual sense for God's eternal present. Of special interest is his conviction that there was no time antecedent to creation; it was God who initiated time along with the rest of creation.

Whitehead, not believing in the reality of points and instants, saw events, which he called actual occasions, as quantized and with temporal duration and spatial extent in the space-time continuum. Time and space are inextricably related. Actual occasions are not predetermined by God but only given a grounding and a goal by God. The absolute uniqueness and unrepeatability of actual occasions clearly implies the irreversibility of time in Whitehead's scheme.

Buber's beautifully poetic and profoundly perceptive expression of the living present provides a steppingstone or taking-off point for the notion of timelessness. His sense of the uniqueness and irreproducibility of the I-Thou meeting implies that time is strictly unidirectional. Also his concepts of I-Thou and I-It will provide a valuable viewpoint from which to examine the aspects of time to be discussed in Part III.

Teilhard de Chardin's view of natural and human evolution proceeding along an axis is clearly grounded in the notion of the irreversible arrow of time. This evolution is seen as directed toward a unified goal which Teilhard terms the Omega Point. He also expresses the opinion that this end will be the Christian end. His goal-oriented picture of evolution interestingly bears some similarities to Whitehead's scheme of goal-influenced actual occasions.

III

Physical and Religious Time Concepts Compared

In this Part the physical and religious time concepts discussed in Parts I and II are compared, with particular attention given to such aspects as the beginning, duration, and end of time; timelessness; the unidirectionality of time; and the interrelation of time and space. Also, the problem of whether the two views of time, physical and religious, constitute some form of an irreducible duality or whether they can somehow be unified is addressed in Chapter 14.

10

Beginnings, Endings, Cycles, Durations

Embedded in many religions throughout the ages is the instinctive, if not spiritual, belief that there was a moment of creation. One of the most powerful expressions of this belief, which enriches the meaning of the Genesis account in the Old Testament, is found in the New Testament: "In the beginning was the Word, and the Word was with God, and the Word was God." (John 1:1) An equally powerful as well as beautiful articulation of this perception is given by Lao-tzu in the *Tao te Ching* (Chapter 25):

> There is a thing inherent and natural
> Which existed before heaven and earth.
> Motionless and fathomless.
> It stands alone and never changes;
> It pervades everywhere and never becomes exhausted.
> It may be regarded as the Mother of the Universe.
> I do not know its name,
> If I am forced to give it a name,
> I will call it Tao, and I name it as supreme.[1]

These Christian and Taoist passages have considerable similarity in general import. Both are saying that there was an ineffable Something in the beginning.* For the Christians it is the Word, which is God; for the Taoists it is the Tao, which, recall, means "Way."†

---

*In the case of the Tao, this is the beginning of a cycle.

†It may also be interesting to note that later in John's Gospel, Jesus says: "I am the *way,* the truth, and the life" (John 14:6).

The virtual universality of what might be termed this creation insight is supported by the fact that creation myths are at the spiritual core of many primitive religions. Indeed there is a rich variety in creation scenarios. There are what C.H. Long terms "emergence" (birth) myths found in the Navaho, Pueblo, and South Pacific cultures. Closely related are the "world-parent" myths of Babylonia, Polynesia, Egypt, and the Zuni. Also myths of creation from chaos or from a cosmic egg are found in Greek, Babylonian, Finnish, and Upanishadic literature, as well as Tahitian folklore. Finally there are the creation-from-nothing ('ex nihilo') myths found in Australian, Mayan, and Maoric traditions among others, not to mention the Judeo-Christian Genesis story.[2]

From this abundant testimony for the existence of a creation insight, let us focus on the Old Testament creation myth in Genesis, as was discussed in Chapter 8. Without doubt it was at least partially on the basis of the Genesis account that St. Augustine (see Chapter 9) arrived at his perceptions of creation and its relation to time. However, his own spiritual insight added enormously to the depth and clarity of the creation event. It will be recalled (Chapter 9) that Augustine stated that God was not in time when he effected creation; he created heaven and earth and along with them, time. There was no time *before* creation (if the use of a time-related word be permitted).

This mature and well-specified religious view of time with respect to creation finds a parallel with the statements of John Wheeler concerning time with respect to the Big Bang. In Chapter 4 we saw that theoretical physicists, using experimental data from both cosmology and elementary particle physics, were able to extrapolate from the present back to the incredibly short time of 10^{-43} sec. after the Big Bang. And in passing back in time there were several stages of unfreezings (or freezings in the forward time direction) at which moments the laws of physical interaction changed. Because of these changes, Wheeler sees this situation as exemplifying what he calls the Law of Mutability or even Law without Law. However, he feels that the ultimate example of Law without Law is the instant of the Big Bang itself, where all physical theory breaks down, along with all our concepts of time and space. Physicists can predict up to a point extremely close to this "wall," but at the wall and beyond, no one knows what occurred.

As we learned in Chapter 4, some physicists, such as Schramm, speculate that time may have existed, for all we know, prior to the Big Bang. Wheeler and others believe not. In any case there seems to be an essential agreement between John Wheeler and St. Augustine in this respect. Although it would undoubtedly require a rigorous essay in physical and religious philosophy to nail down the precise nature of the agreement, I am convinced that on the basic idea of a beginning event antecedent to which there was no time, the two men are saying the same thing in broad outline. This implies also that they are in general alignment concerning the reality of timelessness, to be discussed in the next chapter.

As explained in Chapter 4, for Wheeler the succession of progressively more refined physical laws, each new one serving as the underpinning for its predecessor, prompted him to suggest that they might form some kind of closed circuit, with the observer-participator completing the closure. In the same vein he wondered if quantum theory could be derived from the perception of genesis depending on observership, and wondered also if all physical laws might be derived from almost nothing. But in turn, one cannot help wonder and speculate whether Wheeler might be groping for some kind of concretized refinement, some materialized expression, of what the Toaist calls Tao, the Hindu Brahman, or the Christian the Word. We are reminded of Trefil's observation cited in Chapter 4 that as one goes back in time approaching the Big Bang, the physical theory progressively becomes one of increased simplicity and beauty.[3] Prior to the Big Bang, was there some form of Ultimate Simplicity?

Beside the broad agreement regarding time and timelessness between St. Augustine and Wheeler, there are further interesting comparisons between Genesis and the Big Bang. In the Genesis account, as well as in many other creation myths, the creation event is closely associated with the emergence or production of light. In Chapter 4 we saw that light in the form of high energy photons was present in the very early stages of the birth of the universe, and that light from atoms became possible at about $T_0 + 10^5$ yrs., still quite early on a cosmic time scale. This intense light emission in the early stages of the universe finds a rather remarkable comparison with the light attendant in the creation scenarios of many ancient traditions.

Another interesting similarity between the biblical account and the history of our universe and planet is the fact that the development of man was reached in stages, i.e., the "Days" of Genesis. By the first Day there was light, by the second earth, then plants, then animals, etc., until on the sixth man was made. While some Christian thinkers may find it useful to think of God's Days as consisting of millions or billions of years, that concern is not really what is of interest here. What is interesting is the fundamental insight that there was an evolutionary process in stages leading to man. Everything was not created all at once.

This may not sound as if it were such a revelatory observation now, but for 1000 B.C. it was quite an insight. In the Introduction it was pointed out that the size of man's brain has not changed significantly in 100,000 years. Although brain size is not necessarily an accurate indication of intelligence, men and women at the time of the early recorded religions may nevertheless have been as intelligent as we are today, at least in terms of instinctive insight. What early people did not possess in rational scientific knowledge they may have made up for in intuitional knowledge, much of which would be spiritual. While the ancients may not have been able to fill in all the complex and refined details, they may have been able through spiritual insight to sense reality and time's place in it in broad outline.

The remarkable thing to be noted here is that so many early religious mystics saw that there was a beginning, often involving light emission, thousands of years before anyone heard of the Big Bang theory. This is specifically true of the originators of the creation myths which were drawn upon for the Genesis history. It is further true of the Old Testament writers who sensed that light was involved and that there was some kind of rough time scale in the stepwise evolutionary process. Finally, it is true of St. Augustine with his mature and specific view of time's beginning and of timelessness, some 1600 years before modern cosmology.

Endings

By the very fact that we are dealing with the rather unpredictable future, the comparisons that can be made between projected physical and religious endings are considerably less clearcut than in the case of beginnings. In a real sense this uncertainty is at the present time even supported by the current state of physical and cosmological theory, which as we saw in Chapter 4 cannot yet determine whether we are in an open or closed universe. It begins to look more and more as if we are at the hairline dividing the two rather different cosmological scenarios. Of course if the universe is closed, then parallels with religious endings might be drawn on the basis of the clear finality of the ultimate big crunch. But also there is the further possibility that for comparisons we could regard the end as occurring in about 5 billion years when the sun will evolve and swell into a Red Giant star swallowing up the earth.

Parallel to the indefiniteness as to the universe's future, open or closed, is an indefiniteness expressed generally by the nebulous and visionary character of the endings described in biblical sources such as Daniel and Revelation. However, if we allow ourselves for a moment to indulge in some imaginative speculation, we might compare with somewhat more definiteness the ending scenarios described in these two apocalypes with the two possible cosmological fates projected for the universe. In the two apocalypses, at the end time a sweeping cosmic scenario is described in which judgment is rendered and the good enjoy eternal redemption in God's Kingdom, while the wicked are condemned to everlasting damnation. In this damnation, fire is a prominent feature. If the universe is closed, "fire" of cosmic proportions would certainly be present during some period in the latter stages before the final universal gravitational collapse. So it may not be too difficult to envision the closed universe as correlating with some sort of ultimate damnation. On the other hand, if the universe is open, then we might ultimately enjoy the virtually everlasting existence as "sentient black clouds" visualized by Dyson in Chapter 4. Such a fate might be imagined as being identified with existence in some form of heavenly eternity.

Returning now to a more realistic level, it is nevertheless clear that in their apocalyptic literature both the Jews and the Christians associate the end of the world with the hope of salvation. Remember also that both the Hindu and Mahayana Buddhist cyclical cosmologies had salvational motivations.* The incredibly draining and boring process of repeated births and deaths and never-ending cosmic cycles was no doubt deliberately presented that way so as to make moksha and oneness with Brahman as well as nirvana a much more attractive prospect.

Therefore the end wished and dreamed for by the Hindus is a unity with Brahman, and for the Buddhists it is a reconciliation with the universe in nirvana. In a similar soteriological vein the Jews wish to join God with the realization of the messianic vision of a New Jerusalem and the Christians with the second coming of Christ. Thus religious endings are generally salvational in nature whether they take the form of a release from endless world cycles or a Day of Judgment. Also compared to beginnings, they are necessarily more nebulous and visionary with indefinite time scales and with rich imagery.† Thus in relating the uncertain picture of religious endings with the current uncertainty given by cosmological evidence as to whether the universe is open or closed, we may be prompted to wonder whether there might be a profound significance, perhaps religious, to the fact that the balance between open and closed is as precarious as it presently appears it may be.

In the case of modern religious thinkers such as Whitehead and Teilhard de Chardin discussed in Chapter 8, the end takes

---

*Recall that the Hindus believe we are presently in the kali yuga (see Chapter 5), the last of four epochs comprising the current mahayuga. The four yugas are characterized by progressively shorter duration and increased moral degradation. The kali, the last and worst, will end in vicious destruction and bestiality.

†It is interesting to note with respect to the Christian end that the Gospel writers as well as St. Paul initially believed that Christ's second coming was to come quite soon (see, for example Mark 13:30). When this did not occur, the early Christians had to readjust their eschatological thinking.

more the form of a goal. Remember Whitehead's actual occasions, each with a subjective aim for which God provides a grounding. The actual occasions are lured to their maximum fulfillment by a God, whose power in the world is based on the power of himself as an ideal. An even more clear-cut goal, and on a cosmic scale, is envisioned by Teilhard, who sees a universe of ever-increasing complexity evolving along an axis directed toward an Omega Point, where, in his view, the essentials of Christian eschatology may be realized. A similar view is expressed by Samuel Alexander, who also sees the universe evolving to ever greater complexities with the final stage being Deity.[4] It is a hierarchy of forms of existence which is like a pyramid with space-time at its base and God at the pinnacle. Thus, although the last two of these more modern views of religious endings may have a soteriological aspect, they are primarily regarded as the ultimate deified "Goal" of an evolutionary process without any specific ending scenarios or time scales. This is a further illustration of the difficulty encountered in drawing concrete parallels in time between physically and religiously perceived endings.

Cycles and Durations

The only reasonably accurate idea we have of religious time durations on a cosmic scale is via some of the more encompassing cyclical cosmologies. Undoubtedly the cosmology that involves the longest time durations is that of the Hindus described in Chapter 5. There we saw that the yuga was the smallest cycle and varied in length from 432,000 to 1,728,000 years. The four yugas comprise a mahayuga of 4,320,000 years, one thousand of which (4.32 billions years) make up a kalpa, a day in the life of Brahma. With a night of the same length and a year of 360 such Brahmic days, a 100-year Brahmic life leads to the incredible figure of 311,040 billion years.

No other religion specifically envisions in its scriptures and soteriological cosmologies such vast time durations. It is remarkable that with some combination of religious insight, coupled with some limited knowledge of astronomy and numbers, the Hindu mystics should have arrived at time durations roughly comparable in scale to the present estimate of the age of the universe, about 18 billion years.

Among the closest competitors in this respect are the Mayans who envisioned three intermeshing cyclic years. One was of 260 days, based on the nine-month pregnancy period; a second was the 365-day year; while the third was a 584-day cycle based on the orbit of the planet Venus. Although their calendar dates back only to 3114 B.C. (less than the Hebraic calendar), the Mayan calculations are projected back 400 million years. Remember also we noted in Chapter 7 that the Chinese monk-astronomer I-Hsing calculated that 96,961,740 years had elapsed since the "Grand Origin."

In any event certainly the larger Hindu cyclic durations perceived of as involving billions of years are comparable to cosmological time scales as we know them today. Not only that, it is still possible that the Hindus could be right and the universe does undergo cycles. If the universe does turn out to be closed, then billions of years from now the present expansion will cease, reverse and ultimately result in a big crunch. But maybe it will re-emerge with another Big Bang. As mentioned in Chapter 4, some cosmology theorists are taking seriously the possibility of a cyclical universe, with each successive cycle having a duration increased by at least a factor of two.[5] We saw also in Chapter 3 that Andrei Sakharov postulated a cyclical universe.

From what origin or source would the regeneration of such a cycle occur? The Hindus tell us it is the indefinable Brahman, and the Taoists say it is the nameless Tao. In groping for a physical correlate to these religious concepts, I prefer to imagine a comparison to some form of Wheeler's quantum foam mentioned in Chapter 2. We might speculate that the universe is like a giant fluctuation in the quantum foam that bursts out, expands, then contracts, and is ultimately swallowed up, only to re-emerge later. Is some kind of primal ether or quantum foam the very first in a series of manifestations of Brahman or the Tao that proceed to a blossoming universe? Or is the foam the origin itself and thus identifiable with Brahman and Tao?

Regardless of what specific form such questions and speculations may take, our interest in them helps us understand how the primal underlying process of emergence from (and return to) "nothing" has haunted the imagination and engaged the study of thinkers, both religious and scientific, for so long. Whether such endeavors have dealt with cycles and durations or with beginnings, some early thinkers, though lacking in scientific sophistication, seemed able to use their spiritual insight along with whatever other intellectual tools they had to paint with broad strokes the more focused physical picture that seems to be developing today.

11

Timelessness

Since the evolution of human consciousness men and women, destined to be victims of time in death, have dreamed of transcending the "talons of time,"[1] or as Eliade puts it, the terror of history.[2] Panikkar has observed that of mankind's three primal limitations—knowledge, space, and time—it is the last that is most discomforting.[3] For many, the notion of timelessness has always presented an attractive prospect in terms of immortality and/or salvation.

Either timelessness or everlasting time is a fundamental attribute of the ultimate deity in most religions throughout the world. In any experience of union with such a deity, whether achieved by meditation, sudden inspiration, or prolonged spiritual striving, there is an attendant and strong timeless quality. This is true of the unitive experience of the great Christian mystics, as well as the Hindus in moksha and the Buddhists in nirvana.

Plato in his *Timaeus* saw the universe as patterned after an eternal Living Being, who wished to fashion it as closely as possible to his pattern. The universe could not be endowed with the eternal perfection of the pattern but was made a moving image of it. In particular the Living Being made "that which we call time an eternal moving image of the eternity* which remains

---

*The word "eternity" according to the definition in Webster's *Third New International Dictionary* can mean endless time, the totality of infinite time, or absolute timelessness. Unless otherwise specified, we will be addressing the timeless aspect of eternity when the word is used in this chapter.

forever at one."[4] In such a sense time itself could be said to be a moving image of a timeless Ideal.

Timelessness "Before" and "After" the Universe

Is this universe, that has now lived some 18 billion years, a gigantic yet dynamic sculpture of a Creator, perfect, ideal, and timeless? Is a hint concerning this question to be found in the Grand Unified Theories (GUT), discussed in Chapter 4, where it was mentioned that there was a progression towards simplicity and beauty as the forces became unified going back in time to the Big Bang?[5] Even though the GUT may currently be under revision, some form of it will probably succeed and similar beauty and simplicity still be revealed. Besides, some progression toward such unity is established already with the recent experimentally confirmed unification of the electromagnetic and the weak force, also discussed in Chapter 4.

Whether or not one graces the backward progression toward simplicity and beauty with some spiritual or idealic significance, there is some evidence, as discussed in Chapters 4 and 10, that timelessness may complete the progression. Time may not have existed prior to the Big Bang. It was discussed in Chapter 10 how John Wheeler and St. Augustine agreed that antecedent to creation there was no time, since time is part of creation. The notion of timelessness is then an obvious and powerful corollary to the premise that there was a beginning of time at creation.

Thus in the physical thought of Wheeler, the philosophic thought of Plato, and the spiritual thought of Augustine, there is a common conception of timelessness. It is there in their most refined and imaginative thinking, transcendent, awesome, and essentially inarticulable. But as in the comparison of creation myths with the Big Bang, it is again remarkable how a perception of timelessness derives from such disparate sources as the 4th-century religious views of Augustine and 20th-century physics views of Wheeler.

Just as timelessness can be associated with creation, it can in principle also be associated with the end of the universe, if there is to be one. As discussed previously, it is uncertain now as to whether the universe is open or closed; but if it is closed, it will end in a "big crunch." At this point time may end and timelessness prevail.

In the last chapter it was noted that the comparison of physical and religious endings is not as definitive as with beginnings. While the Judeo-Christian endings treated in Chapters 8 and 10 include eternity as an essential component, what biblical eternity means is often not very clear. It is sometimes thought to signify everlasting time. If this is the case, then the biblical endings can only have an obvious comparison to the open universe scenario. On the other hand, if timelessness is intended, then the apparent parallel would be with a closed universe. But, as indicated earlier, these endings are primarily judgmental and salvational in character. Thus a comparison in terms of a literal physical ending appears difficult to make.

Cosmic Cycles and Timelessness

However, the sense of timelessness associated with endings is quite definite when we look at the completion of a Hindu cosmic cycle. It is then that the universe, along with time and space, returns to the timeless, ineffable Brahman. Then timelessness would result, just as it may with a closed universe in modern cosmological theory. Furthermore, as noted in Chapter 10, if the universe were not only closed but cyclical as suggested by some cosmologists, the parallel between the Hindu and modern cosmology would be quite close. In both cases there would be a "hiatus" of timelessness between cycles.

A generally similar but somewhat more diffuse comparison can be made with the concept of return in Taoist tradition. Just as all things return to Brahman, so also they return to the Tao. Therefore the combination of cyclicality and timelessness is evident also in the Taoist cosmology, which again could be compared with the cyclical closed universe hypothesis in current physical theory.

Another important aspect of the timelessness characteristic of both Brahman and the Tao is the implication that they are perpetually pervasive, yet detached, transcendent, and outside time and space while the universe is undergoing its cycles. That is, it is a timelessness that exists independently of the occurrence of beginnings and endings. Furthermore, it is this transcendent timelessness that is available in moksha and nirvana when the endless cycle of births, lives, and deaths, as well as the cosmic cycles, are left behind.

As suggested in the last chapter, such pervasive yet transcendent properties prompt one to wonder whether the timelessness of the Tao or Brahman can in some way be identified with the timelessness of the submicroscopic world, Wheeler's quantum foam for example. Remember, in this realm of dimensions of the order of 10^{-33} cm. and thus hidden by the Heisenberg Uncertainty Principle, time has no meaning, according to Wheeler.

Regardless of these speculative comparisons, nowhere in either the literature of physics or religion is the sense of timelessness so beautifully and supremely delineated as in the Vedantic Hindu concept of Brahman. The accumulated wisdom of millennia in the Hindu religious tradition lays claim to the reality of timelessness. It increasingly appears that it may be one of the missions of physics to refine the nature and import of something approximating this concept.

Finally, it should be observed that in general there is a timeless aspect to any kind of cyclical temporal system. As Schlegel has pointed out, one can only sense cyclical and linear temporality when both exist.[6] That is, one can only sense that one is involved in cyclical time if linear time also exists as a reference. If we were totally immersed in cyclical time with no distinction from one cycle to another, we would be in what amounts to a timeless state. To be in such a state was probably an underlying purpose of archaic man in yearly reliving a primal, sacred event, so that he could deal with the terror of history.

Timelessness and Reversible Time

There is also a timeless quality to the concept of reversible time in the physical microscopic world. Because of the complete symmetry in the mathematical equations describing virtually all microscopic phenomena, these phenomena can be adequately described by time proceeding in either the plus or minus direction. In effect from the viewpoint of the macroscopic world, microscopic time does not make any difference and is regarded by many physicists as a mathematical parameter. This lends a static, timeless quality to microscopic reversible time. Schroedinger once observed that a mathematical truth is timeless.[7] It is as if the mathematical equations of microscopic physics may exist as a kind of eternal Platonic ideal, or as one of Whitehead's eternal objects discussed in Chapter 9.

Perhaps the most interesting parallel to microscopic time symmetry can be found in Buddhist meditative thought. At a stage in the meditative experience, time may sometimes be seen in one dynamic sweep in which past and future are symmetrically viewed from the timeless, unfettered eye of the living present. I believe this can be inferred, for example, from the words of Lama Govinda quoted at the end of Chapter 6. That time can at least in a meditative sense be reversed and thus symmetrized was evident in the Buddhist yogic practice, also discussed in Chapter 6, wherein the yogin was able to pass back in time to the cosmic beginning (i.e. initiation of a cycle) and thus effect an "emergence from time." Therefore, I feel that some sense of time symmetry can be extracted from the Buddhist yogic experience that broadly parallels the essential notion of time symmetry (and its related timeless aspects) found in the microscopic world of physics.

The Living Present and Timelessness

In discussing Martin Buber's powerful and perceptive insights into the nature of the I-Thou dialogue, it was noted that the central ingredient in such a relationship was the ability to step totally into the living present. Under such conditions all sense of time is lost, it is an essentially timeless experience. It was also observed that such a state must at some level be a precursor in any path to a lasting unitive spiritual state, such as that undergone by the Christian mystics or as reached in moksha or nirvana. The sense of total awareness and sensitivity to the living moment is especially apparent in the Buddhist meditative tradition, and such meditation is deliberately designed as an aid in achieving nirvana and its attendant timelessness.

St. Augustine speaks of God's "eternal now."[8] It is only in this "now" that we can have access to God. In the totality of the vibrant and timeless present, we may with grace experience full communion with the timeless God. A reverence for the living moment is thus indispensable for any sense of timelessness; the living moment is the looking glass through which timelessness can be perceived. It is in the eternal now that apprehensions of time and timelessness find their interface.

The Judeo-Christian tradition tells us that it was this same timeless God who created the world. In the biblical creation process the first thing created was light: "And God said, Let there be light, and there was light" (Gen. 1:3). As cited earlier, light is also the initiating feature of many other creation myths. Thus light is often sensed to be among the closest physical attributes that can characterize God or any deity. Light also is often a strong feature of any experience of spiritual illumination or near-death vision.

On the other hand, physically light in its particle mode consists of photons, as explained in Chapter 2. Photons in the sense of special relativity are timeless or atemporal, as Fraser terms it.[9] That is, as implied in Chapter 1, if a transformation is made to a system moving at the velocity of light (the photon's velocity), the clocks in that frame of reference are slowed to zero. God in his atemporality has somehow provided an atemporal object, the photon, as a very early feature of creation, as a means of communication for man, and, perhaps in some cases, as a manifestation of his (God's) presence. Thus a consideration of light, its primality in creation, and its essential physical atemporality affords a fascinating and provocative connection between religious and physical views of timelessness.

Let us conclude this chapter with a well-known remark relevant to timelessness, made by Einstein, who was largely responsible for our modern view of the physical properties of light. Though he often stated that he did not believe in an anthropomorphic or personal God, he had a profound reverence for the beauty and order in a universe governed by Reason. And his apprehension of timelessness was quite apparent in the comment he made at the death of his dear and life-long friend, Michele Besso:

> And now he has preceded me briefly in bidding farewell to this strange world. This signifies nothing. For us who are convinced physicists, the distinction between past, present, and future is only an illusion, even if a stubborn one.[10]

12

The Arrow of Time

This now-famous expression coined by Eddington beautifully captures the irreversibility and unidirectionality of time.[1] Unidirectional flow is generally sensed by two different modes of ordering events as to time. The first is establishing a temporal sequence through cognizance of "before" and "after." Whether the events are in the past or future, such an ordering is in principle possible through the use of memory or anticipation. The second mode is forming such a sequence through perception of "past," "present," and "future."

It may seem superficially that there is little to distinguish the two modes, but they are usually used in different contexts. The first is most often in a more objective and relatively static context, because past events can be ordered with respect to before and after by the memory and future events so ordered by knowledgeable anticipation. The second mode is in a more dynamic and subjective context because it is distinguished by the constantly advancing "knife edge" of the present, personally experienced and continually sharply separating the past from the future.[2]

Another important consideration that seems inherent in the understanding of the arrow of time is the fact that we only notice change with respect to something that is permanent or relatively permanent.[3] We see ourselves aging while the house we live in, or the county courthouse in the center of town seems to stay the same. Also it was observed earlier how linear time is sensed relative to cyclical time.[4] For example, there is a certain linear progressiveness to the continued and reliable repetition of the seasons as we grow older (i.e., the seasons serve as a clock).

Still another aspect of time's arrow that often seems to characterize it, especially from a subjective viewpoint, is its anticipatory nature. This is most powerfully expressed by the philosopher Husserl and later refined and clarified by Merleau-Ponty.[5] The latter holds that "one's world is carried forward by lines of intentionality which trace out in advance at least the style of what is to come."[6] That is, this philosopher sees time not so much as an advancing line but a *"network of intentionalities."* In this thinking he follows Husserl, presenting the situation diagramatically as shown in Figure 7. As one proceeds in time from A to B left to right along the line of present moments, moment A recedes into the memory (by dropping down into the region of memory indicated in the figure) and becomes A′, a retention of the memory. Proceeding from B to C, moment B similarly becomes the memory retention B′, and A′ recedes further into the memory as A″. The vertical lines in the figure are therefore those of retention or memory.[7] The slanted lines are lines of intentionality, dotted toward the future, what Merleau-Ponty terms lines of "protention," or perhaps more simply lines of anticipation. The lines of anticipation or intentionality thus schematically represent the distinctive view of the arrow of time expressed by this philosophy. It is characterized and distinguished by an anticipatory outlook or a continuous state of presupposition. It is almost as if the anticipation, intentionality, or presupposition were some kind of subjective propagator for the arrow of time.

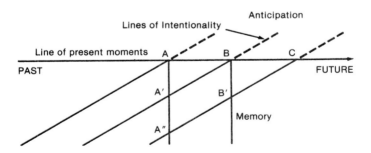

Figure 7. From M. Merleau-Ponty, Phenomenology of Perception. *Atlantic Highlands, NJ: Humanities Press, 1962, p. 417. By permission of the publisher.*

Physical and Religious Arrows: Comparisons and Parallels

Keeping in mind these few selected notions about perception and sensing time's arrow, let us examine physical and religious arrows and how they compare. In Chapter 3 at least three different physical arrows were discussed: the thermodynamic, the historical, and the cosmological. The thermodynamic arrow had as its basis the irrevocable increase of entropy or disorder. The historical had as its basis just the opposite, the progressive growth of order or information. At the root of the cosmological arrow was the expansion of the universe, which has been in progress for the last 18 billion years and shows few signs of slowing.

There are two fundamental properties that all three arrows share. The first is that they all describe time as a linear phenomenon, or more elegantly as a progressive linear sequence of temporal states of being. The second is that this linear time is unidirectional or irreversible, i.e., the above sequence of states proceeds in only one direction along the temporal line and never reverses. Such apparently obvious statements seem trivial and would actually be so but for the fact that reversible time is so predominant in the microscopic world and that cyclical time dominates several religions to this day.

However, from a physical point of view irreversible time is vital in the macroscopic world, for without it there would be no progression to higher order, which includes humans. There would also be no progression to greater disorder, which is also very important in the behavior of the natural world around us.

The foregoing features of linear irreversible time, which are an irreducible common property of all three of the physical arrows of time, likewise underlie the views of irreversible time to be found in religious thought. The notion of a religious arrow of time is especially intense in the thinking of Teilhard de Chardin, Samuel Alexander, and Alfred North Whitehead. Indeed this is true of the Judeo-Christian concept of time as a whole. This is because the arrows characterizing these three thinkers and Judeo-Christian tradition are value-endowed. They are either pointing to some ultimate and divine goal or are being continually and divinely influenced in their progress.

In the case of Teilhard linear irreversible time is inherent in his conception of the evolution of the universe along a purposeful axis, as described in Chapter 9.[8] Each stage of evolution provides the grounding for the breakthrough to the next greater complexity and order. In the last stage the axis of evolution finds itself aligned with the Christian purpose, the axis being pointed to an ultimate goal wherein Teilhard's "Omega Point" becomes identified with the consummation of God's will. This clear implication of inherently directed linear time is likewise to be found in the approach of Samuel Alexander, who also saw "Deity" as the target toward which the evolutionary arrow is pointing.

Whitehead's concept of actual occasions progressing in a value-endowed process presents a somewhat different picture than that of Teilhard.[9] This is because the value is continually impregnating an actual occasion through its subjective aim, for which, recall, God provides a grounding (see Chapter 9). Indeed it is via God's conjunctive influence and interaction with actual occasions that he is in a sense part of the process that Whitehead sees the universe as undergoing. Thus for Whitehead, God is not so much a goal to which an evolutionary axis is pointing as a continual immanent and transcendent influence on the evolutionary process, luring the actual occasions of this process to their maximum fulfillment. There is also no question as to the basic irreversibility of time in Whitehead's system. The unquestioned uniqueness and unrepeatability of each actual occasion insures the progress of time's arrow.

The last example of a value-endowed arrow of time mentioned above is found in the concept of time developed in the Judeo-Christian tradition. In Chapter 8 it was shown how the people of Israel, by experiencing a whole series of sacred events wherein Yahweh intervened in history, gradually developed a linear sense of history, and ultimately of time. This was of course inherited later by the Christians. In addition, for both Jews and Christians a sense of linearity undoubtedly derived from the belief in a creation that was not repeated and in some form of final Judgment Day, whether the coming of the Messiah for the Jews or Christ's second coming for the Christians.

This sense of linearity was then strongly influenced by Jewish and Christian eschatology. For example, in the case of the Christians, if we can hazard some kind of median position among the

New Testament scholars, it was a realized eschatology, because it was felt that the Kingdom of God is already here and available to those who would seek it with complete and sincere diligence. But it was also an eschatology unfulfilled in its totality until Christ's second coming. Hence we see the curious mixture of present and future. This notion of a future-oriented present suggests a general correlation with the view of Husserl and Merleau-Ponty discussed earlier, that sees time as progressing along lines of intentionality that are clearly future-directed.

In any event it is this proleptic eschatology and the divine goal of God's Day of Redemption and Judgment that justifies calling the Christian arrow value-endowed. It is guided by the light of the future and directed toward Judgment Day. In essence a similar picture is seen in the case of the Judaic arrow. There is always somewhere deep in the heart of every true Jewish worshipper the hope and anticipation of the Messianic Kingdom in some form. Thus the Judaic and Christian arrows are following similar paths in basic essentials.

While all of the above value-endowed arrows have in common with the physical arrows mentioned earlier (thermodynamic, historical, cosmological) the fundamental properties of linearity and unidirectionality, perhaps a more detailed comparison of physical and religious arrows is possible here. It would certainly seem that among the three physical arrows the above value-endowed religious arrows might be most closely identified with the historical arrow. The progressive development of order and/or information characterizing this arrow seems to imply a growth to greater beauty, wisdom, or excellence. Thus it appears most consistent with a value-endowed arrow either pointed toward or influenced by divine truth.

However, a correlation with the historical arrow is not as clear-cut in the case of the linear irreversible time implied in Buber's thought (see Chapter 9). It will be recalled that it was from Buber's concept of the complete uniqueness of each I-Thou meeting, never to happen again in exactly the same way, that one can imply linear irreversible time. However, there is no direct expression of value endowment, at least in the sense of the religious arrows thus far discussed.

On the other hand, there may be value implied in the sense of the profound worth to be found in stepping wholly into the living moment in an unbridled I-Thou relation. It is, at least in part, through such acts of courage in the face of fearful vulnerability, and through answering to the very deepest in oneself, that a person can become "that for which God intended."[10] Yet in the true I-Thou relation, according to Buber, anticipation does not and cannot come into play. It would only dilute the total sense of presence and be inconsistent with letting the dialogue happen. Therefore the arrow of time that can be drawn from Buber's thought does not have a goal-oriented or influenced characteristic except in the sense, at the ultimately personal level, of daring to be the person God intended. Therefore the Buberian arrow can only be identified with the historical arrow of the physical realm in a qualified way.

Another example of a view of time that does not seem to be as goal-directed or influenced is the Chinese linear time discussed in Chapter 7. There it was stated that Chinese time has both linear and cyclical aspects. In the general area of social and family relations and hierarchy, time is linear. The *I Ching (Book of Changes)* tells us that each moment can be denoted by a number indicative of the quality of that moment. Therefore, while there is in a real sense a value placed on the moments of Chinese linear time, it is not obviously goal-directed or influenced. Hence, this time also may not be easily identified with the physical world's historical arrow.

Therefore, all of the arrows of time discussed, whether physical (thermodynamic, historical, cosmological) or religious (Teilhard, Alexander, Whitehead, Judeo-Christian, Buber, Chinese), have in common a fundamental linearity which is additionally unidirectional or irreversible. However, it appears that a closer correlation specifically with the physical historical arrow can best be made in the case of Teilhard, Whitehead, Alexander, and the Judeo-Christians.

A final very interesting observation concerning the irreversibility of time can be drawn from an analogy with Whitehead's thought. It will be remembered that he claimed that a limitation of God is his goodness, otherwise God would be responsible for the evil in the world. It would appear that another limitation of God may be the irreversibility of time at the macroscopic level. God does not reverse Time.* Either he cannot or chooses not to allow the beautiful symmetry in time that is manifested in almost the entire microscopic world to be directly experienced at the macroscopic level.

---

* A conceivable challenge to this statement might come from Buddhists who, in a certain yogic mode of consciousness, pass back in time to the world's beginning, as discussed in Chapter 6.

13

Interdependence of Time and Space

In those rare and precious moments when we wholly open ourselves to our vibrant surroundings, time and space take on a different appearance and seem to reveal a deep essence of integration. The living "now" assumes an immanent, timeless quality; we are in tune with the Eternal Now. Our surroundings, our space become the universe in the sense that we are totally at peace with our place in it and feel a full relation to it, all of it. In effect, we are a microcosm of the universe. All space and time are distilled into the "here" and "now."

Rare though these moments are, it is through such experiences that one can best understand the validity of the words of the mystics and religious contemplatives from both East and West. It is certainly in the context of such moments, or meditative states of consciousness based thereon, that the Buddhist thinkers mentioned in Chapter 6 were able to make their observations about the interrelation of time and space. This interrelation was expressed there with compelling beauty and clarity in quotations from the renowned Eastern scholar D. T. Suzuki and the great Tibetan mystic Lama Govinda. In both cases the universe was viewed from a meditative state as a living continuum in which time and space were interpenetrating and integrated.

A fundamentally similar view of the interdependence of time and space was expressed by Whitehead, as discussed in Chapter 9. Again recall that he saw the universe as a process of actual occasions, having a subjective aim for which God provided a grounding. It was God also who "lured" the actual occasions to their optimal fulfillment. These events were regarded as quanta of experience occupying some finite duration and extent in the space-time continuum. For Whitehead there was no time without space and no space without time; they become commingled in the concresence of an actual occasion.

In his book *Space, Time, and Deity* the incisive philosophic and religious thinker Samuel Alexander represented a closely related perception of time and space. He convincingly argued that time as such could not exist without space and vice versa, and went so far as to insist that "space is in its very nature temporal and time spatial."[1] The intimate relation of time and space was imaginatively described using the human body as a metaphor: time is the mind of space and space the body of time. In contrast to Whitehead, Alexander expressed himself in terms of points and instants and viewed time and space in a world that was a complex of point-instants, known as "pure events."[2] Nevertheless, there is considerable similarity in the views of Alexander and Whitehead with respect to space-time interdependence.

The essential features of the foregoing religious and philosophic perceptions of the interdependence of time and space find a comfortable comparison with similar aspects of relativity and quantum theory. In relativity (see Chapter 1) the interconnectedness of time and space was most apparent in relating measurements in one frame of reference to those in another frame moving at some velocity relative to the first. In Chapter 1 it was explained that the space measurements in Andrea's frame depend not only on the space measurements in Bob's but also on the time measurement in his. Similarly, Andrea's time measurement depends not only on Bob's time measurement but also in general on the space measurements in his frame.

Via light with its great but finite velocity, this inextricable interconnection of time and space relates all parts of the universe with all other parts, so that the physical laws experienced locally by any one part are identical to those experienced by any other. In this dynamic universe all the particles are moving relative to each other, often at speeds approaching that of light, and their relationship to one another entails such an interdependence of space and time. Thus space and time are interrelated in just such a way that it is possible for each of us, even if moving at great relative speeds, to see nature, including light itself, in exactly the same way. If it were not that way, the universe would be chaos.

A consideration of the transformation equations treated in Chapter 1 also indicated that our perception of simultaneity has to be enlarged. Two events seen as simultaneous in one frame of reference are observed at different times in another; our notion of simultaneity has been relativized. The relativization is generally consistent with the view of the Buddhist mystics who, as noted above, see time and space as interpenetrating with an attendant softening of notions of simultaneity.

Another relativistic phenomenon, this time arising from gravitational effects described by general relativity and discussed in Chapter 1, was the relationship between mass and time. In the proximity of a large mass, a clock will slow down, so that a gravitational dilation of time occurs. But from Chapter 1 we also know that, according to general relativity, mass determines the shape or curvature of space. Time and space are then affected by mass, and it is thus through mass that they are somehow at least indirectly linked.

A relation between mass and time is also implied in quantum theory through the Heisenberg Uncertainty Principle as it applies to the simultaneous measurement of time and energy. Recall that a precise measurement of energy required a complementary imprecise knowledge of the time, and vice versa. The complementary relationship between time and energy can be regarded, via Einstein's $E = mc^2$ equation, as such a relationship between mass and time. Therefore the Uncertainty Principle can be applied to time and mass measurements. Such an application of the principle comes into play particularly in the measurement of the mass of the very short-lived elementary particles, most of which have been observed in the last thirty-five years. The shorter the lifetime of such an unstable particle, the more uncertain is the determination of its mass. This is another way in which time and mass are related.

From the foregoing remarks we have seen views concerning the interaction of time and space from the thought of Whitehead, Alexander, and the Buddhist mystics. The interdependence of time and space is also revealed in effects arising from the relativity and quantum theories. From a broad qualitative viewpoint it may be said that in essence similar views are being expressed by the religious (or religious philosophic) sources and modern physics.

However, here is another instance where more penetrating religious and physical philosophy than can be given in a book of this scope may be needed to fix more accurately where and to what degree the different views of space-time relation are aligned. Actually some of the groundwork for such study, some essentials of which were distilled and presented in this book, may already be available in the works of Whitehead and Alexander.* Additionally, it is fascinating to realize that some perceptions of the interdependence of space and time were perceived by the Buddhist sages centuries ago and that these perceptions broadly summarize what is found in modern physics.

The Interpenetration of Matter

We learned above that matter in some fashion serves as a link between time and space. Thus an examination of the interpenetration of matter may help highlight the general concept of interpenetration and aid in our view of it as applied to time and space. In the survey of quantum theory in Chapter 2, it was learned that all matter, in particular all microscopic matter such as electrons, protons, atoms, and molecules, could be characterized by a probability function or pattern. This pattern often assumes a bell-shaped form such as depicted in Figure 3a. The probability is greatest that the particle is at the center of the pattern where the curve is maximum. Each side of the pattern trails off to very low probabilities. However, the probability that the particle might be located at those extremes, though very slight, is nevertheless not zero, even out to the far reaches of the universe.

It is in this probabilistic sense, and only in this sense as we saw in Chapter 2, that all matter, all particles of matter are interpenetrating. Nevertheless this form of interpenetration finds a conceptual parallel with Mahayana Buddhist thought that sees all things as mutually pervading and interactive. The message of the *Avatamsaka sutra* as expressed by Buddhist scholar Eliot tells us:

---

*For further study of Whitehead the reader should consult the references given in Chapter 9, especially Hammerschmidt, *Whitehead's Philosophy of Time*. All volumes of Alexander's *Space, Time, and Deity* should also be consulted.

> In the heaven of Indra there is said to be a network of pearls so arranged that if you look at one you see all others reflected in it. In the same way each object in the world is not merely itself but involves every other object and in fact is everything else.[3]

A similar perception is implied in the words of D.T. Suzuki:

> When the one is set against all the others, the one is seen as pervading them all and at the same time embracing them all in itself.[4]

These two quotations clearly transmit the universal sense of interpenetration that characterizes the view of the Mahayana mystics and finds such a fascinating parallel in quantum theory.

Although we must take care to realize that quantum theory is saying that there is only an infinitesimal probability that a local object or particle will be found at a distant location, the idea nevertheless haunts the imagination. The philosophic and spiritual import of this feature of the quantum theory is still a matter of rich controversy today. Since we as observers, or participators as Wheeler puts it, are also made of matter, the related problem of observer theory also is affected by our concept of interpenetration. That is, where do the limits of the observer's domain end and those of the object observed begin? Another related and imaginative question is how do thought, insight, and emotion operating in the human brain affect the atoms and molecules therein, which in turn through interpenetration relate to all matter in the external world.[5]

The Quantization of Time and Space

A question that continues to engage the attention of some theoretical physicists is whether time and space are quantized, i.e., do they exist as infinitesimal, discrete, and fundamental units or granules? For example, the time units have often been called "chronons," although as was noted in Chapter 2, the prevailing current theoretical opinion does not give much support to this concept. Nevertheless, it has not been conclusively disproven and variations of it still appear in the literature. A related concept forwarded by Nobel laureate T.D. Lee was described at the end of Chapter 2. There it was stated that on the reasonable fundamental assumption that only a finite number of measurements can be taken in a given volume of space, Lee developed what he terms a discrete mechanics as opposed to the conventional continuous mechanics. The four-dimensional space-time continuum is looked upon as having a kind of lattice-like structure. Although Lee does not specifically embrace the notion of space and time quantization in his mathematics, his system bears a close conceptual relationship to such a notion.

However, Whitehead in his impressive metaphysical system embracing both religious and physical concepts, supports the general idea of space-time quantization. In his system, discussed earlier in this chapter and in Chapter 9, he finds no use for the concept of points in space and instants (points) in time. As we have seen, actual occasions are quanta of experience which require some extension in space-time for their concrescence or fruition; hence space-time in some sense is quantized. Therefore there is some correlation between the space-time models proposed by some theoretical physicists, who are "believers" in quantization, and the religiously influenced metaphysics of Whitehead. Although the subject as a whole is not currently receiving much attention, it is nevertheless not dead and could possibly enjoy more support at some future time.

Extreme Concentrations of Space-Time

We learned in Chapter 4 that, as one traces back in time toward the instant of the Big Bang, the laws of physics change at certain critical stages. At these stages there was a simplification in the form of a unification of one of the forces of nature (e.g. electromagnetic, gravity, nuclear, etc.) with another. However, going back in time beyond about 10^{-43} seconds after the Big Bang (or $T_0 + 10^{-43}$ seconds as expressed in Chapter 4), the mathematics breaks down, along with our concept of space and time. In this progressive simplification, do space and time finally become totally mixed beyond 10^{-43} seconds before finally going out of existence? Wheeler and St. Augustine tell us that time and space did not exist prior to creation. This may be true. However, as cited earlier, others disagree or feel that nothing can be said about the state of the universe antecedent to the Big Bang. But if time and space did have a prior existence, in what form was it? Does the enormous concentration of matter and energy characterizing the Big Bang mean that space and time were compressed into some form of unity at the instant of their birth? The quest for answers continues and probably always will.

A fascinating religious parallel to the physical view of such space-time concentration is again found in the words of Lama Govinda. He speaks of

> ...a living continuum in which time and space are integrated into that ultimate incommensurable 'point-like' unity, which in Tibetan is called 'thig-le' (Sanskrit 'bindu'). This word which has many meanings, like 'point, dot, zero, drop, germ, semen, etc.' occupies an important place in the terminology and practice of meditation. It signifies the concentrative starting point in the unfoldment of 'inner space' in meditation, as also the last point of its ultimate integration. It is the point from which inner and outer space have their origin and in which they become one again.[6]

The correlation of this description with that associated with the Big Bang should be apparent. In each case the integration of space and time is associated with a strong concentration, a point-like unity. Does the accomplished Buddhist meditator in his most profound state of consciousness become the ultimate participator in John Wheeler's sense? In this state where space and time are integrated and concentrated analogous to the Big Bang, is the practitioner coming near to closing the circuit of the successively underpinned laws of nature (otherwise apparently terminated by the Big Bang) as described by Wheeler in Chapter 4?

14

Time: A Duality or Unity?

One of the fundamental problems inherent in any attempt to comprehend something as elusive as time is that we are so inextricably and intimately interrelated with it in the natural universe. Time, along with ourselves, is *in* nature; nature is not in time. That is, the prevailing current opinion, whether physical, philosophic, or religious, seems to be that time does not exist "out there" beyond the universe, continually flowing on and on. It is one of the entities that is part of the universe, as we ourselves are.

The whole situation appears to be an example of an axiomatic principle that seems to operate in many areas of study. It goes something like this: Complete knowledge of a system cannot be acquired if the observer is part of the system. We cannot get outside the universe to determine fully the nature of time in the universe and our deep enmeshment with it. Accordingly, we cannot claim absolute objectivity in any view of time, or indeed in anything else. It is therefore far better, instead of uselessly straining for absolute objectivity, to use what objectivity is available in conjunction with what one hopes is an enlightened subjectivity in approaching a problem.

Such an approach is consistent with the prime motivation of this book, namely, to examine time jointly from the viewpoint of physics, the most extreme of the objective, rational disciplines, and of religion, the most subjective and intuitive discipline. Being restricted to what might be termed a limited rational-intuitive or objective-subjective approach, which is dual, it is not surprising that time would have, or appear to have, a duality. This duality has already been expressed in different forms, a few of which were briefly mentioned in the Introduction. Let us survey some of these forms in order to obtain a more comprehensive perspective of what the duality is that time exhibits and perhaps also to learn more about the nature of time itself, as well as time as a potential unity.

Time as a Duality

Different names have geen given to modes assumed by the fundamental rational-intuitive duality with respect to time, and these have been elaborated with considerable refinement. In the Introduction, Eliade's concept of sacred and profane time was touched on briefly, and aspects of it were further developed in Chapter 8.[1] The sacred time was regarded by archaic peoples as the only real time, the time when the cosmogonic creation event was re-enacted with full living force. It was a time of complete purification, redemption, and rebirth, a time of blessing by and communion with the mythical hero-god. Thus this sacred time was one in which the intuition was given free reign and emotional and spiritual response was total.

The time between these events, profane time, was that lived pursuing the everyday tasks and chores that are one's lot. It, of course, was the time when the primitives undoubtedly brought to bear all the rationality they could muster in effectively performing these tasks. So that it is probably safe to suppose that they utilized some elementary analysis in their view of time in doing such chores.

It seems sensible to correlate sacred and profane with the more general categories of intuitive and rational. However, it is somewhat more difficult to find such a parallel with the more specialized physical and religious (or spiritual) categories, because, while Eliade's sacred time can easily be identified as a spiritual view of time, time involved with daily tasks cannot be specifically identified with physical time. Nevertheless, the approximate validity of a general comparison of the dualities seems apparent.

Similar comparisons can be made with Buber's companion concepts of I-Thou and I-It. Again in the I-Thou experience spirituality, intuition, and instinct are in full flower. Whether the relation is with a man, a tree, a poem, or God, one is living in the present time with free and complete grace. The vibrant present and the benign flow of indistinguishable living moments is seen with a total subjectivity. On the other hand, the I-It experience connotes in a sense a withdrawal from the living present. It entails a distancing from and objectification of whatever or whomever the experiencer is relating to. Rational analysis and logical thinking best come into play here.

The states of I-Thou and I-It find an easy comparison, not only with the general duality of intuitive (subjective) and rational (objective) time concepts, but also with the more specific duality of religiospiritual and physical. To put it another way, one's perception of time when in the midst of an I-Thou experience is largely of a spiritual nature, while in the I-It state rationality, objectivity, and analysis charactertistic of physical observation come into play. Therefore in my view identifying time in the I-Thou state with religiospiritual time and in the I-It with physical time is supportable, despite the fact that undoubtedly many a physicist has undergone an I-Thou experience in the laboratory or at his desk devising a theory.

Also alluded to in the Introduction was Park's division of time into two kinds, which he calls Time 1 and Time 2.[2] In essence Time 1 is the time of physical theory that is represented in the equations of physics and registered on a clock dial. Time 2 is the time of human consciousness, primarily characterized by being grounded in the living present. An example of the contrast between the two times can be found in music, wherein Time 1 is the time of musical composition and structure while Time 2 is the time of rhythm. The visual arts find their expression usually in Time 2. Contemplative, analytic study in science and history uses Time 1. Thus Park's two times are tantamount to another expression of the general classification discussed above, namely, rational, objective (Time 1) and intuitive, subjective (Time 2). Religiospiritual time would obviously be a specialization of Time 2.

Denbigh in his thorough and perceptive work posits three concepts of time: 1) the time of conscious awareness; 2) the time of theoretical physics; 3) the time of thermodynamics and the evolutionary sciences such as biology.[3] Park, using a more general framework, would probably label the first of these as Time 2 and the last two as Time 1.

It should be evident that common to all of the time duality concepts presented is the general classification into the rational-objective and intuitive-subjective. At the extremes of this duality we find the most sensitive and profound of the intuitional pursuits, the religiospiritual time view, and the most exquisitely detailed of rational pursuits, the time view of mathematical physics.

Complementarity and Yin-Yang

Is there a way of somehow relating or connecting the components of this duality, at least partially depolarizing them, or finding some threads of implicit unity? One approach already suggested in the Introduction is through the use of a generalization of Bohr's Principle of Complementarity. This principle, as discussed in Chapter 2, was conceived by Bohr for applications to the particle-wave duality in quantum theory. However, as indicated earlier, it has been generalized by many thinkers and applied to a variety of dichotomies in other fields.[4]

There are certain fundamental criteria that an entity that presents itself to us as a duality must fulfill in order for us to apply the generalized complementarity principle to it. First, we must be certain that it is indeed a single phenomenon or entity that we are observing. Second, the phenomenon or entity must appear differently to the two modes of observation. Third, the two modes must be independent and mutually exclusive. That is, one mode cannot be reduced to the other or subsumed by the other. Neither mode gives a complete description; both modes are necessary for completeness. This implies, of course, that the two modes are limited; but it also implies that from some more transcendent viewpoint apparently unavailable to us, a complete picture may be achievable in one sweeping observation.

Just as Bohr attempted to attain some degree of unity in the wave-particle duality through the use of complementarity, it is interesting to explore the application of a more generalized version of the principle to the general duality existing between the intuitive (subjective) and rational (objective) views of time. Also the principle may be applicable to the more specific duality of religiospiritual and physical concepts of time. I believe that for both the general and the more specific duality, a case can be made for the above criteria for complementarity being satisfied. That is, one phenomenon is being observed; neither mode of observation is complete; and one cannot be reduced to or subsumed by the other.

The generalized complementarity principle finds an interesting and beautiful comparison with the Chinese yin-yang principle. For the Chinese these two complementary realities are the basis of operation for all things in the world. In Chapter 7 we saw that yin represents the passive, feminine, and intuitive aspect of things, while yang corresponds to the active, masculine, and rational aspect of things. Therefore, the notion of complementarity is instinctive and deep-seated in the East and is not new. Indeed Bohr became aware of this after a visit to China and found so much beauty in and identity with the yin-yang principle that he often used the yin-yang symbol for complementarity (see Figure 5). For time comparisons yin would obviously be associated with intuitive (subjective) or religiospiritual time and yang with rational (objective) or physical time.

At first sight it may seem contradictory to claim on the one hand that there may be some validity to the application of generalized complementarity to physical and religious time concepts as a duality, and on the other to maintain (as I have in several instances in Part III) that the two viewpoints give similar perceptions of time. In the first place this application of complementarity is much more general with some attendant diffuseness in definition; that is, here the two aspects cannot be defined with logic as rigorous as in the case of the wave-particle dualism. Second and more importantly, just because two modes of observation give complementary and mutually exclusive views of a phenomenon does not mean that they cannot say some of the same things about it. For example, in the case of the wave-particle duality with a photon, both modes agree that the phenomenon is a radiation of zero rest mass; both give it the same frequency; both the same velocity and the same energy. Furthermore, it must be remembered that the complementarity principle was introduced as a possible method of demonstrating some unity, not disparity in the two time views.

Other Possibilities for Unity

Another way that some thinkers have hypothesized about time in the quest for unity is that it may be multidimensional. Just as space is three dimensional, perhaps time has two or more dimensions. For example, are the two views of time, the rational and the intuitive discussed above, somehow merely a unity with two dimensions?

Samuel Alexander postulated three dimensions for time, which were not physically based but philosophically based dimensions.[5] They were what he termed: successiveness, irreversibility, and transivity. The intricate philosophic arguments he uses to justify such a division are out of place here. Also, according to Benjamin it appears that more work may need to be done to develop fully Alexander's concept.[6] Dunne also devised a multidimensional model of time, but from the viewpoint of time as seen in psychic or paranormal states of consciousness.[7]

Some have speculated on the possibility that time might somehow be represented by what is known mathematically as a complex number. Such a number has a real component, which is the ordinary number we see in daily use, and what is called an imaginary component. Here the word imaginary is not used in its conventional sense but has a specific mathematical definition.* There is a whole branch of mathematics devoted to mathematical functions using such complex numbers as variables. In fact, in certain parts of the mathematical formalism of special relativity, it is useful to treat time as an imaginary number so that a self-consistent four-dimensional space-time framework can be formulated. Time is also used as an imaginary number in certain cases in the field of chemical kinetics.

In any case, probably one reason why the idea of time as a

---

*An imaginary number always has a factor of (i.e. is multiplied by) $\sqrt{-1}$ (the square root of -1) which is usually denoted by the symbol "i." Thus the complex number $3 + 4i$ can be plotted as a point in a 2-dimensional complex plane, 3 units along the "real" axis and 4 along the "imaginary" axis.

two-dimensional or two-component quantity is hard to take seriously is that the idea is just as difficult to envision as four-dimensional space would be.* Nevertheless, as was cited in Chapter 4, physicists are working on Supergravity theories with as many as eleven space-time dimensions. That is, in addition to the four conventional ones, there may be seven other "compactified" dimensions, representing unobservably minute regions of the physical continuum (perhaps similar to Wheeler's minimicroscopic "quantum foam"). Moreover, recently Nobel laureate Andrei Sakharov has published a paper which takes seriously the possibility of multidimensional time.[8] He postulates that in addition to our single ordinary macroscopic time there exists an even number (2, 4, 6, or . . .) of time dimensions, again characteristic of apparently unobservable compactified regions.

Park believes that possibly there may be some connection between his Time 1 and Time 2 which might be realized through a deeper understanding of the human brain. His suggestion may have some merit in view of all that has been learned in recent decades about the interactive operation of the left and right hemispheres of the brain, although he does not mention this in his book. Generally the left hemisphere performs more analytic and rational tasks while the right handles more intuitive and instinctive responses. This seems at least partially consistent with evidence which indicates that the left brain is dominant in distinguishing the temporal order of two closely timed events, while the right brain possesses a better sense of time durations as well as velocity and accelerations.[9] This is a fertile area of continuing active study in experimental psychology, and additional interesting results should be forthcoming.

Probably the most serious recent attempt to find an underlying unity in time concepts is that made by Fraser in his recent monograph *The Genesis and Evolution of Time*.[10] He sees time as having a natural history or evolution, just as other entities including ourselves. He proposes what he calls the "principle of

---

*An interesting argument for there being only one temporal dimension given by Fraser is that otherwise there would be no guarantee for the continuing identity of any material object. (See J.T. Fraser, *The Genesis and Evolution of Time*.)

temporal levels," based on the idea that at each significant level or stage in the universe's evolution to ever more complex phenomena, a distinct and unique temporality or kind of time was manifested. He further postulates that these temporalities "coexist in a hierarchically nested, dynamic unity," which is still developing and open ended. As each temporality has evolved from the previous, from which it received its grounding, it continues to exist in interaction with its forebears in such a nested hierarchy. It is a unique and ingenious concept that will undoubtedly provoke discussion both in scientific and philosophic circles for a considerable time.

Another possibility for unity may perhaps be found through a deeper examination of the notion of timelessness. Is timelessness the link or grounding that can relate intuitive to rational time or religio-spiritual to physical time? We have seen on several occasions how some apprehension of timelessness seems attendant in any spiritual experience of being totally in the living present. In addition there are axioms and equations used in physics which are timeless truths. The Pythagorean theorem, though learned initially in plane geometry, enjoys an extensive universal use throughout physics; it is a timeless rational truth.

Thus there is some form of relationship to or grounding in timelessness found in both kinds of time. Perhaps then Taoist and Hindu thought should be taken even more seriously, because it is these traditions that see nature, and time with it, cyclically emerging and then returning to the timeless Tao or Brahman. Are both kinds of time intimately related to and simply manifestations of some timeless reality? This is one of the most profound questions presented to humankind, and some sense of the challenge it presents is expressed by T.S. Eliot: "But to apprehend the point of intersection of the timeless with time, is an occupation for a saint."[11]

The foregoing remarks may at some point have prompted another question which bears strongly on the matter of time as a unified concept. That is, from a more carefully focused viewpoint just what is the nature of the two perceptions of time, whether thought of as an intuitive-rational duality or the more specific spiritual-physical duality? Let us first explore one side of these dualities by first asking the question: Is it only the intuitional, subjective, or religiospiritual time that is fundamentally credible to us? On consideration, is not virtually all of rational, analytical, physical, and objective time somehow rendered static, abstracted, idealized, or spatialized?

The physical time we rationally think about and measure is recorded on a clock dial or a digital readout; or recorded in terms of some fixed array of molecules in the memory bank at a location in our brain or a computer; or set down on a piece of paper in some mathematical equation. All these are spatial in one way or another. The living present that we subjectively and spiritually experience is for most of us the truest and deepest reality we know. An acorn falling from an oak tree can have its time of fall recorded, its trajectory of fall analyzed, and its impact on the earth measured. But to one in the living "now" its reality is expressed as a living event in a dynamic and vibrant universe. In Longfellow's words:

> What is time? The shadow of the dial, the striking of the clock, the running of the sand, day and night, summer and winter, months, years, centuries—these are but arbitrary and outward signs, the measures of Time, not Time itself. Time is the life of the soul.[12]

Perhaps Whitehead[13] is right when he faults physical scientists for not using the human senses in a much more direct fashion to describe adequately the nature around us. Are our intuitional and spiritual notions of time, Park's Time 2, or Eliade's sacred time, the true primal apprehension of time available to us? Are rational time and physical time then an extrapolation from this experienced time and find their value in all of the fascinating and practically useful detailing in terms of abstract equations and physical models? Does not the fact that there are parallels, similarities, or correlations in the religious and physical time concepts as shown in this book actually support such a view? Can our notion of time derived from intuition and spirituality therefore be likened to the trunk and roots of a tree, while our rational and physical views correspond to its branches and leaves?[14]

Let us now explore the other side of the issue and pose questions highlighting the case for the primacy of the rational, physical concept of time. Is not our intuitive, spiritual view of time to a considerable extent a matter of individual and unique personal experience that would never be identical in a total or absolute sense to the experience of others, though it may bear some similarity? A fundamental criterion for any rational, and in particular any scientific, endeavor is that the method's logic and/or observation be clearly specified and be such that they can be performed by anyone properly equipped (either mentally or with instruments). Further, the results of such logic and/or observation should be capable of being reproduced by anyone using the same methods. Therefore, is it not true that we can only find a universally agreed upon concept of time, independent of individual human bias, through a rational, scientific approach?

Cannot a physical concept of time describe the temporal behavior of each of the atoms that contribute to the material composition of our bodies? Indeed, does not a physical concept of time also describe the temporal behavior of an incredibly vast array of natural phenomena throughout a universe whose size makes us seem exceedingly insignificant? It may be obvious that the question of which view of time is the most fundamental and real is very unsettled and will continue indefinitely to engage the thought of philosophers, physicists, biologists, psychologists, and religious scholars.

From what has been discussed in this chapter, it would seem that in some respects time exhibits itself as a duality and in others a unity. Given the present state of our knowledge and development, I believe that there probably are two different views of time, whether classified as subjective and objective, intuitive and rational, or religiospiritual and physical. However, I think that the two views find a degree of unity by virtue of some form of generalized complementarity. A total picture of time would not be possible without both views. I also believe that there is further evidence for unity to be found in some of the aspects of time treated in this book. That is, in the case of beginnings of time, time durations, timelessness, and especially time's arrow and space-time interrelation, the religious sources in many respects have painted in broad strokes what the modern physicists have photographed in much finer detail.

Unity Through Time as a Link

To give us some clue to resolving the paradox of time, I have often wondered if what the world needs is another unique person, perhaps someone like St. Augustine, who has found a mystical union with God or Reality (e.g., moksha or nirvana), and yet who is well trained in rational disciplines, e.g., physical science. It would certainly seem that someone deeply immersed in and conversant with both the physical and spiritual worlds might be able to say something definitive about the mystery of time. Perhaps such a one can help place the apparent duality of time in its true perspective, and can give us a fuller understanding of Panikkar's provocative but unexplained remark: "Time is at the crossing point between consciousness and matter."[15]

This observation bears some similarity to the thought of Denbigh, who not only sees time generally as a relational concept connecting events, but more specifically as a link between the mental and material aspects of the brain.[16] A related view of time as a link was expressed by Eddington and quoted at the conclusion of the Introduction: "In any attempt to bridge the domains of experience belonging to the spiritual and physical sides of our nature, time occupies the key position."[17]

It is as if we ourselves through a much more intimate integration of our spiritual and physical selves might be able to sense time as the fluid link between the two, and thus concurrently see time as a unity, and even get a glimpse of the meaning of timelessness. No longer under the illusion that time is something we can grasp, inspect, and register, by allowing ourselves to ride free on its arrow, we might identify with it and its vibrant integrating presence in ourselves. Realizing that we are somehow linked by time to our ancestors and will be to our progeny, perhaps we may even extend our perception to see time as some kind of cohesive connection with the rest of the universe and indeed to the moment of its creation, via that awesome limiting vehicle, the velocity of light. Perhaps this universal time linkage may be perceived as implicit in Fraser's concept, mentioned earlier, of time as evolving through a nested, open-ended series of temporal levels embracing in a hierarchical unity progressively more complex phenomona.[18] But regardless of how it may be perceived, a deeper understanding of the notion of time as a link may lead to a clearer view of time as a unity. This engaging conjectural line of thought suggests but one of many possible paths for study for which this book may perhaps serve as a first step.

Epilogue

According to a colleague who was visiting a university in Nova Scotia, he saw there on a bathroom wall a piece of graffiti which went something like this: "Nature [or God] provided us with time so that everything would not happen at once." Aside from possibly being construed, perhaps with some justification, as a testimony to the quality of the intellectual level of the institution, it is a profound statement. With the universe God or Nature has opened up, perhaps out of what seems a point of nothing, a vast system of space, time, and matter to *be* in. Time has been granted us to understand our being in it and our true relation to its Origin. For man and woman it is especially time with its intrinsic vitality that is the means, the divine link, by which an interaction with God or Nature can be known. It is the telescope through which the Divine can be seen.

In the Shanidar Cave in Iraq floral pollen was detected in the grave of a *homo sapiens neanderthalensis* who was buried 60,000 years ago. This is considered strong evidence that flowers were a part of the burial ritual at that time. Thus this brief opening, this fleeting view, in time and space that is afforded us has been treated with deep reverence for millennia. It is my belief that only when the problem of time is approached with such reverence will it ever be fully understood. In the words of von Schiller, "Time is man's angel."*

---

*Quoted in *Voices of Time*, J. T. Fraser, ed. (Amherst: University of Massachusetts Press, 1981), p. lvii.

Notes

Introduction

1. M. Eliade, *The Myth of the Eternal Return* (Princeton: Princeton University Press, 1971).
2. D. Park, *Image of Eternity* (Amherst: University of Massachusetts Press, 1980).
3. K. G. Denbigh, *Three Concepts of Time* (New York: Springer-Verlag, 1981).
4. F. Capra, *The Tao of Physics* (New York: Bantam Books, 1977); R. Jastrow, *God and the Astronomers* (New York: Warner Books, 1980); M. Talbot, *Mysticism and the New Physics* (New York: Bantam Books, 1981).
5. I. G. Barbour, *Issues in Science and Religion* (Englewood Cliffs, NJ: Prentice-Hall, 1966); K. Cauthen, *Science, Secularization and God* (New York: Abingdon Press, 1969); H. N. Wieman, *Religious Experience and the Scientific Method* (Carbondale, IL: Southern Illinois University, 1954); A. R. Peacocke, ed., *The Sciences and Theology in the Twentieth Century* (Notre Dame, IN: University of Notre Dame Press, 1981); R. W. Burhoe, *Toward a Scientific Theology* (Ottawa, Canada: Christian Journals Limited, 1981).
6. N. Bohr in *Nature,* Vol. 121, 1928, p. 580; also in *Physical Review,* Vol. 58, 1935, p. 696.
7. A. S. Eddington, *The Nature of the Physical World* (New York: Macmillan Inc., 1929).

Chapter 1

1. A. Einstein, *Relativity, The Special and General Theory* (London: Metheun, 1920); R. P. Feynman, R. B. Leighton, and M. Sands, *The Feynman Lectures on Physics,* Ch. 15-16 (New York: Addison-Wesley, Inc., 1965); P. C. W. Davies, *Space and Time in the Modern Universe*

(Cambridge: Cambridge University Press, 1977); C. W. Misner, K. S. Thorne, and J. A. Wheeler, *Gravitation* (San Francisco: W. H. Freeman and Co., 1973); A. P. French, *Special Relativity* (New York: W. W. Norton and Co., 1968); R. Resnick, *Introduction to Special Relativity* (New York: Wiley, 1968); E. F. Taylor and J. A. Wheeler, *Spacetime Physics* (San Francisco: W. H. Freeman and Co., 1966).

2. From Newton's "Principia" as set down in E. Marsh, *Science of Mechanics* (London: Open Court Publishing Co., 1907), p. 226.

3. Michelson and Moreley, *Silliman Journal,* Vol. 34, 1887, pp. 333-427; Morley and Michelson, *Philosophical Magazine,* Vol. 24, 1887, p. 24.

4. In A. Einstein, A. H. Lorentz, H. Weyl, and H. Minkowski, *The Principle of Relativity* (New York: Dover Publications, 1973) p.75.

5. Davies, *Space and Time in the Modern Universe.*

6. Denbigh, *Three Concepts of Time,* p. 46, quoting C. D. Broad.

7. John A. Wheeler, *Physics and Austerity: Law without Law* (Austin: University of Texas, 1982), p. 65.

8. See J. T. Fraser, ed., *Voices of Time* (Amherst: University of Massachusetts Press, 1981), p. 475.

9. See O. Costa de Beauregard, "Time in Relativity Theory: Arguments for a Philosophy of Being" in J.T. Fraser, ed., *Voices of Time* (Amherst: University of Massachusetts Press, 1981), p. 417.

10. Ibid.

11. See M. Capek, in *Voices of Time,* p. 434.

12. *Bulletin de la Societe Francaise de Philosophe,* April, 1922, p. 108.

13. E. Meyerson, *La Deduction Relativiste,* p. 100-104.

14. Fraser, *Voices of Time,* p. 473.

Chapter 2

1. R. T. Weidner and R. L. Sells, *Elementary Modern Physics* (Boston: Allyn and Bacon, Inc., 1965); Feynman, Leighton, and Sands, *The Feynman Lectures on Physics;* E. Merzbacher, *Quantum Mechanics* (New York: John Wiley and Sons, Inc., 1961); R. Resnick, *Basic Concepts in Relativity and Early Quantum Theory* (New York: John Wiley and Sons, 1972); R. Eisberg and R. Resnick, *Quantum Physics of Atoms, Molecules, Solids, Nuclei, and Particles* (New York: John Wiley and Sons, 1974); L. I. Schiff, *Quantum Mechanics* (New York: McGraw-Hill, 1968).

2. Weidner and Sells, *Elementary Modern Physics,* p. 142ff.

3. *Zygon, Journal of Religion and Science* (Chicago: Chicago University Press, 1964-1980).

4. See D. Bohm, *Wholeness and the Implicate Order* (London: Routledge and Kegan Paul, 1980).
5. J. A. Wheeler, "Genesis and Observership," Princeton University Report, 1976, p. 9.
6. J. A. Wheeler, "Physics and Austerity: Law without Law," University of Texas Report, 1982, p. 13.
7. W. Heisenberg, *Physics and Philosophy,* (New York: Harper and Row, 1958).
8. Wheeler, *Physics and Austerity,* p. 8ff.
9. Ibid.
10. B. S. DeWitt, "Quantum Gravity," in *Scientific American,* Dec. 1983, p. 112.
11. Wheeler, *Physics and Austerity,* p. 58.
12. T. D. Lee, "Time as a Dynamical Variable and Discrete Mechanics" in *Bulletin of the American Physical Society,* Vol. 28, 1983, p. 700; and "Can Time be a Discrete Dynamical Variable?" in *Physics Letters,* Vol. 122 B, 1983, p. 217.
13. Feynman, Leighton, and Sands, *The Feynman Lectures on Physics.*

Chapter 3

1. D. Park, *Image of Eternity* (Amherst: University of Massachusetts Press, 1980).
2. E. Schroedinger, *Mind and Matter* (Cambridge: Cambridge University Press, 1967).
3. R. Schlegel, *Time and the Physical World* (New York: Dover Publications, Inc., 1968), P. 17ff.
4. Eddington, *The Nature of the Physical World,* Chapter 4.
5. Schlegel, *Time and the Physical World.*
6. D. Layzer, "The Arrow of Time," in *Scientific American,* Vol. 256, 1975, p. 56.
7. A. D. Sakharov, "Many-sheeted Models of the Universe", *Soviet Physics JETP,* Vol. 56, 1982, p. 705.
8. S. Weinberg, *The First Three Minutes* (New York: Bantam Books, 1977).
9. F. Wilczek, "The Cosmic Asymmetry between Matter and Antimatter," in *Scientific American,* Vol. 243, 1980, p. 82.
10. I. Prigogine, *From Being to Becoming* (San Francisco: W. H. Freeman and Company, 1980), p. 49.
11. Ibid., p. 159.
12. Ibid., p. xiii.
13. Ibid., p. 176.
14. Ibid., p. 196.

Chapter 4

1. Wheeler, *Physics and Austerity*, p. 65.
2. G. J. Whitrow in *Voices of Time*, p. 572.
3. *Physics Today*, May, 1983, p. 17.
4. Ibid.
5. A. H. Guth, *Physical Review*, Vol. D23 (1981), p. 347.
6. Weinberg, *The First Three Minutes*.
7. R. Jastrow, *Until the Sun Dies* (New York: W. W. Norton and Co., Inc., 1977); also *God and the Astronomers* (New York: Warner Books, Inc., 1980).
8. D. N. Schramm, "The Early Universe and High Energy Physics," in *Physics Today*, April, 1983, p. 27; J. Trefil, "How the Universe Began" in *Smithsonian*, May, 1983, p. 32.
9. Ibid.
10. Schramm, "The Early Universe and High Energy Physics."
11. Trefil, "How the Universe Began."
12. J. A. Wheeler, "Beyond the Black Hole," Princeton University Report, 1979.
13. D. A. Dicus, J. R. LeTaw, D. C. Teplitz, and V. L Teplitz, *Scientific American*, Vol. 248, March, 1983, p. 91.
14. F. J. Dyson, "Time without End: Physics and Biology in an Open Universe," *Reviews of Modern Physics*, Vol. 51, July, 1979, p. 447.
15. P. A. M. Dirac, "The Large Numbers Hypothesis and the Einstein Theory of Gravitation," Proceedings of the Royal Society, London, Vol. A365, 1979, p. 19.
16. Dyson, "Time Without End."
17. R. Jastrow, *The Enchanted Loom* (New York: Simon and Schuster, 1981).
18. Dicus, LeTaw, Teplitz and Teplitz, *Scientific American*, p. 98.
19. Ibid., p. 100.
20. B. Carter, in *Scientific American*, December, 1981, p. 160.
21. Collins and Hawking, as cited in *Scientific American*, December, 1981, p. 168.
22. J. A. Wheeler, "Genesis and Observership," *Center for Theoretical Physics Report*, University of Texas, Austin, 1976, p. 1.
23. Ibid., p. 30.
24. R. H. Dicke, "Dirac's Cosmology and Mach's Principle," *Nature* 192, p. 440-1.
25. Wheeler, "Genesis and Observership," p. 41.
26. Ibid., p. 29.
27. Wheeler, "Physics and Austerity," p. 42.
28. Wheeler, "Genesis and Observership," p. 60.
29. Wheeler, "Genesis and Observership," p. 71.

30. A. Velenkin, "Creation of Universes from Nothing," *Physics Letters,* Vol. 177B, 1982, p. 25.
31. Schramm, "The Early Universe and High Energy Physics."

Chapter 5

1. T. J. Hopkins, *The Hindu Religious Tradition* (Belmont, CA: Dickenson Publishing Co., 1971), p. 25.
2. Ibid., p. 33.
3. Ibid., p. 36.
4. Ibid., p. 38.
5. Ibid., p. 54.
6. R. H. Robinson, *The Buddhist Religious Tradition* (Belmont, CA: Dickenson Publishing Co., 1970), p. 41.
7. Swami Nikhilananda, tr., *The Upanishads* (New York: Harper and Row, 1963), p. 121.
8. E. Deutsch, tr. *The Bhagavad Gita* (New York: Holt, Rinehart and Winston, 1968).
9. Ibid., p. 39.
10. Ibid., p. 42.
11. Ibid., p. 86.
12. Shankara, Swami Prabhavananda and Christopher Isherwood, trs., *Crest-Jewel of Discrimination* (New York: New American Library, Mentor Books, 1970).
13. Deutsch, *The Bhagavad Gita,* p. 83.
14. Nikhilananda, *The Upanishads,* p. 131.
15. Hopkins, *The Hindu Religious Tradition,* p. 101.
16. Ibid., p. 279.
17. M. Eliade, *Yoga, Immortality and Freedom* (Princeton: Princeton University Press, 1958), p. 79.
18. A. Balslev, *Study of Time in Indian Philosophy* (Wiesbaden: Otto Harvassowitz, 1983).
19. R. Reyna, *Introduction to Indian Philosophy* (Bombay: Tata McGraw-Hill, 1971), p. 202.
20. M. Hiriyanna, *Outlines of Indian Philosophy* (London: George Allan and Unwin Ltd., 1932), p. 324.
21. Ibid., p. 404.
22. Shankara, *Crest-Jewel of Discrimination,* p. 13.
23. Ibid., p. 38.
24. Quoted in Nikhilananda, *The Upanishads,* p. 20.

Chapter 6

1. R. H. Robinson, *The Buddhist Religion* (Belmont, CA: Dickenson Publishing Co., Inc., 1970), pp. 13-15.

2. Ibid., p. 20
3. E. Conze, *Buddhist Thought in India* (Ann Arbor: University of Michigan Press, 1967), p. 71.
4. W. S. Rahula, *What the Buddha Taught* (New York: Grove Press, Inc., 1959), p. 35.
5. Reyna, *Introduction to Indian Philosophy*, p. 106.
6. R. Kloetzli, *Buddhist Cosmology* (New Delhi: Motilal Banarsidass, 1983).
7. Eliade, *The Myth of the Eternal Return*, p. 115.
8. A. B. Keith, *Buddhist Philosophy in India and Ceylon* (New York: Gordon Press, 1974), p. 163 ff.
9. Ibid.
10. Eliade, *Yoga, Immortality, and Freedom*, p. 184.
11. H. Zimmer, *Philosophies of India* (Princeton: Princeton University Press, 1951), p. 62.
12. Eliade, *Yoga, Immortality, and Freedom*.
13. D. T. Suzuki in Introduction to B. L. Suzuki, "Mahayana Buddhism," The Buddhist Lodge, London, 1938, p. xxvii; also quoted in F. Capra, *The Tao of Physics* (New York: Bantam Books, 1977).
14. Lama Anagarika Govinda, *Foundations of Tibetan Mysticism* (York Beach, ME: Samuel Weiser Inc., 1969), p. 116 ff; also quoted in F. Capra, *The Tao of Physics*.

Chapter 7

1. H. Nakamura, *Ways of Thinking of the Eastern Peoples* (Honolulu: East-West Center Press, 1964), pp. 157, 234.
2. Ibid., pp. 107, 202.
3. J. Needham, "Time and Knowledge in China and the West," in *Voices of Time*, p. 101.
4. Nakamura, *Ways of Thinking of the Eastern Peoples*, pp. 167 and 284.
5. M. von Franz, *Time: Rhythm and Repose* (London: Thameson and Hudson, 1978), p. 20.
6. Needham in *Voices of Time*, pp. 100, 117.
7. Ibid., p. 133ff.
8. Ibid., p. 111.
9. H. Welch, *Taoism, The Parting of the Way* (Boston: Beacon Press, 1957), p. 2.
10. Ibid., p. 3.
11. Ibid., p. 20.
12. Ibid., p. 21.
13. Ibid., p. 21.
14. Ibid., p. 23.

15. Ibid., p. 42.
16. Ibid., p. 36.
17. Chang Chung-yuan, tr. *Tao: A New Way of Thinking* (New York: Harper and Row, 1975), p. 27.
18. Welch, *Taoism, The Parting of the Way*, p. 55.
19. Ibid., p. 68.
20. Needham in *Voices of Time*, p. 98.
21. Welch, *Taoism, The Parting of the Way*, p. 16.
22. Ibid., p. 21.

Chapter 8

1. Eliade, *The Myth of the Eternal Return*, pp. 21, 35.
2. Ibid., p. 52.
3. Ibid., p. 36.
4. Ibid., p. 150.
5. Ibid., p. 81.
6. Ibid., p. 90.
7. Ibid., p. 21.
8. Ibid., p. 5.
9. Ibid., p. 23.
10. Ibid., p. 63.
11. G. von Rad, *The Message of the Prophets* (New York: Harper and Row, 1965), p. 81.
12. Ibid., p. 83.
13. Ibid., p. 78.
14. G. Delling in R. Kittel, ed., *Theological Dictionary of the New Testament*, Vol. 3 (Grand Rapids, MI: William B. Eerdmans Publishing Co., 1965), p. 455.
15. Ibid.
16. See J. Barr, *Biblical Words for Time* (Naperville, IL: Alec R. Allenson, Inc., 1962). See also A. Momigliano, "Time in Ancient Historiography" in *History and Theory* Vol. 5, Beiheft 6, 1966, p. 1.
17. von Rad, *The Message of the Prophets*, p. 86.
18. Ibid., p. 84.
19. Eliade, *The Myth of the Eternal Return*, p. 160.
20. Ibid., p. 108-110.
21. J. L. McKenzie, *Dictionary of the Bible* (New York: Macmillan Publishing Co., Inc., 1965), p. 157.
22. Ibid., p. 158.
23. O. Cullman, F. V. Filson, tr. *Christ and Time* (Philadelphia: Westminster Press, 1950), p. 81ff.
24. J. Marsh, *The Fullness of Time* (New York: Harper and Brothers, 1952), p. 174ff.

25. C. H. Dodd, *Parables of the Kingdom* (New York: Scribner, 1961) pp. 46, 49, 197.
26. W. G. Kummel, *Promise and Fulfillment* (London: SCM Press Ltd., 1957), pp. 16, 143.
27. Cullman, *Christ and Time,* p. 84.
28. C. M. Laymon, ed. *The Interpreters One-Volume Commentary on the Bible* (New York: Abingdon Press, 1971), p. 437.
29. Ibid., p. 947

Chapter 9

1. St. Augustine, *Confessions* R. S. Pine-Coffin, tr., (Baltimore: Penguin Books, 1961), p. 253.
2. Ibid., p. 264.
3. Ibid., p. 274.
4. Ibid., p. 276.
5. K. Thompson, *Whitehead's Philosophy of Religion* (The Hague: Mouton and Co., 1971), p. 22.
6. R. B. Mellert, *What is Process Theology* (New York: Paulist Press, 1975), p. 22.
7. Ibid.
8. Ibid.
9. Thompson, *Whitehead's Philosophy of Religion,* p. 23.
10. Ibid., p. 23.
11. W. W. Hammerschmidt, *Whitehead's Philosophy of Time* (New York: Russell and Russell, 1947), p. 83.
12. A. N. Whitehead, *Process and Reality* (New York: Harper and Row, 1929), p. 343.
13. A. N. Whitehead, *Religion in the Making* (New York: Macmillan Co., 1926), p. 153.
14. Ibid., pp. 153 and 156.
15. Ibid., p. 154.
16. Thompson, *Whitehead's Philosophy of Religion,* p. 58.
17. Ibid., p. 59.
18. Whitehead, *Process and Reality,* p. 161.
19. A. N. Whitehead, *Adventures of Ideas* (New York: Macmillan Co., 1929), p. 213.
20. Whitehead, *Religion in the Making,* p. 95-8.
21. Thompson, *Whitehead's Philosophy of Religion,* p. 183
22. Whitehead, *Process and Reality,* p. 413.
23. L. Bright, *Whitehead's Philosophy of Physics* (New York: Sheed and Ward, 1958), p. 43.
24. A. N. Whitehead, *Science and the Modern World* (New York: Macmillan Co., 1925), p. 114.

25. Hammerschmidt, *Whitehead's Philosophy of Time*, p. 26-7.
26. Ibid., p. 18.
27. A. N. Whitehead, *An Enquiry Concerning the Principles of Natural Knowledge* (London: Cambridge University Press, 1919), p. 7.
28. Hammerschmidt, *Whitehead's Philosophy of Time*, p. 54.
29. A. N. Whitehead, *Modes of Thought* (London: Cambridge University Press, 1938), p. 79.
30. Hammerschmidt, *Whitehead's Philosophy of Time*, p. 4.
31. M. Buber, *The Origin and Meaning of Hasidism*, M. Friedman, tr. (New York: Horizon Press, 1960).
32. M. Buber, *I and Thou*, R. G. Smith, tr. (New York: Charles Scribner's Sons, 1958); also translated by W. Kaufman, (New York: Charles Scribner's Sons, 1970).
33. Buber, Smith tr., *I and Thou*, p. 11.
34. Ibid., p. 11.
35. Ibid., p. 24.
36. Ibid., p. 34.
37. M. S. Friedman, *Martin Buber, the Life of Dialogue* (New York: Harper and Bros., 1955), p. 116; M. Buber, *Between Man and Man* (New York: Harper and Bros., 1965), P. 140-1.
38. Ibid.
39. Ibid.
40. Buber, *I and Thou*, p. 12.
41. Ibid., p. 13.
42. Buber, *Between Man and Man*.
43. J. L. Russell, "Time in Christian Thought" in *Voices of Time*, p. 74.
44. Ibid., p. 75.
45. P. Teilhard de Chardin, "Turmoil or Genesis?," originally published in *L'Anthropologie,* Paris, 1947. Translated in a chapter in Ian Barbour, *Science and Religion* (New York: Harper and Row, 1968), p. 210.
46. Ibid., p. 217.
47. Ibid., p. 217.
48. Ibid., p. 220.
49. Ibid., p. 222.
50. Russell in *Voices of Time,* p. 76.

Chapter 10

1. Chu Ta-kao, tr., *Tao te Ching* (Boston: Mandala Books, 1982).
2. C. H. Long, *Alpha, The Myths of Creation* (New York: George Brazilier, 1963).
3. Trefil, "How the Universe Began" in *Smithsonian*, p. 32.

4. S. Alexander, *Space, Time and Deity* (New York: The Humanities Press, 1920).
5. Dicus, LeTaw, Teplitz, and Teplitz, *Scientific American,* p. 100.

Chapter 11

1. This expression is due to Dr. Robert Russell, Center for Theology and Natural Sciences, Berkeley, CA, 1982.
2. Eliade, *The Myth of the Eternal Return,* p. 160.
3. R. Panikkar, "Time and Sacrifice—The Sacrifice of Time and the Ritual of Modernity" in Fraser et al, ed., *The Study of Time III* (New York: Springer-Verlag, 1978), p. 687.
4. Plato's "Timaeus" as quoted in Park, *Image of Eternity,* p. 101.
5. Trefil, "How the Universe Began," p. 32.
6. Schlegel, *Time and the Physical World,* p. 14.
7. Schroedinger, *What is Life?* and *Mind and Matter,* p. 154.
8. St. Augustine, *Confessions.*
9. Fraser, *Voices of Time,* p. 473ff.
10. B. Hoffmann, *Albert Einstein, Creator and Rebel* (New York: Plume Books, 1972), p. 258.

Chapter 12

1. Eddington, *The Nature of the Physical World.*
2. A. C. Benjamin in *Voices in Time,* pp. 6, 7.
3. Ibid.
4. Schlegel, *Time and the Physical World,* p. 14.
5. M. Merleau-Ponty, *Phenomenology of Perception* (Atlantic Highlands, NJ: Humanities Press, 1962), p. 410ff.
6. Ibid.
7. Ibid.
8. Teilhard de Chardin in *Science and Religion.*
9. See Whitehead, *Process and Reality, Religion in the Making,* and *Adventures of Ideas.*
10. M. Buber, *Good and Evil* (New York: Charles Scribner and Sons, 1952), p. 142.

Chapter 13

1. S. Alexander, *Space Time, and Deity,* Vol 1 (Atlantic Highlands, NJ: Humanities Press, 1920), p. 44.
2. Ibid., p. 48; and A. C. Benjamin in *Voices of Time,* pp. 27, 28.
3. C. Eliot, *Japanese Buddhism* (London: Edward Arnold and Co., 1935), p. 109; also quoted in F. Capra, *The Tao of Physics.*

4. D. T. Suzuki, *The Essence of Buddhism* (Tokyo: Hozokan, 1968), p. 52; also quoted in F. Capra, *The Tao of Physics.*
5. R. Toben and F. A. Wolf, *Space, Time, and Beyond* (New York: Bantam Books, 1983), pp. 36, 37.
6. Lama Anagarika Govinda, *Foundations of Tibetan Mysticism,* p. 116ff.

Chapter 14

1. Eliade, *The Myth of the Eternal Return.*
2. Park, *The Image of Eternity,* p. 100ff.
3. Denbigh, *Three Concepts of Time.*
4. I. G. Barbour, *Myths, Models, and Paradigms* (New York: Harper and Row, 1974), p. 75ff.
5. Alexander, *Space, Time, and Deity.*
6. A. C. Benjamin, "Ideas of Time in the History of Philosophy," in *Voices of Time,* p. 28.
7. J. W. Dunne, *An Experiment with Time* (New York: Macmillan Company, 1927).
8. A. D. Sakharov, "Cosmological Transitions with Changes in the Signature of the Metric," *Zhurnal Eksperimentalnoi u Teoreticheskoi Fisiki,* Vol. 87, 1984, p. 375.
9. R. L. Atkinson, R. C. Atkinson, and E. Hilgard, *Introduction to Psychology* (New York: Harcourt, Brace, and Jovanovich, 1983), p. 47ff; also Prof. Gerry Levi, Department of Psychology, University of Chicago, private communication, 1984.
10. J. T. Fraser, *The Genesis and Evolution of Time* (Amherst: University of Massachusetts Press, 1981), pp. 1-lx.
11. As quoted in *Voices of Time,* pp. 1-lx.
12. Ibid.
13. A. N. Whitehead, *The Concept of Nature* (London: Cambridge University Press, 1920), pp. 29, 45-6.
14. This analogy is due to F. Capra, *The Tao of Physics.*
15. Panikkar, "Time and Sacrifice—the Sacrifice of Time and the Ritual of Modernity," in *The Study of Time III,* p. 684.
16. Denbigh, *Three Concepts of Time.*
17. Eddington, *The Nature of The Physical World.*
18. Fraser, *The Genesis and Evolution of Time.*

Index

Abraham, 105, 108
Acceleration, 20
Alexander, S., 141, 152-53, 155, 158-60, 169
Anthropic Principle, 67-68, 72
Anatman, 87
Apocalyptic, 114
Aranyakas, 76
Arjuna, 77-78
Asamkhyeya, 89
Ashoka, 77
Atman, 78, 87
Atom, size of, 25
Avatamsaka school, 91, 160-61

Bhagavad Gita, 77-78, 80
Big Bang, 37, 39, 50, 68-72, 136-37, 145; Theory, 55-59, 138, 162-63
Black body radiation, 26, 58
Black holes, 63, 66-67. See also Heisenberg, W.
Bodhisattva, 89
Bohr, N., 25, 36; Complementarity Principle, 6, 29, 167-68
Boltzmann, L., 44-45, 52
Brahma, 80-81, 141
Brahman, 76-84, 100, 137, 140-42, 146-47, 171; Nirguna, 80; Saguna, 80
Brahmanas, 76
Buber, M., 117, 124-27, 130; I-It concept, 126-27, 165-66; I-Thou concept, 125-27, 148, 154-55, 165-66

Buddha, Gautama, 85-87, 89
Buddhism, 77, 85-86, 94-95; cosmology, 89-92, 140; sects, 87-88; view of time, 87-93, 147-48, 159-61

Canaan, 105-7; Canaanite calendar, 106; myths, 110-11
Carter, B., 68
Christ, Jesus, 109; ministry, 109, 111; resurrection, 109, 111; second coming, 110, 140, 153-54
Complementarity Principle, *See* Bohr, N.
Conditioned, definition of, 86; Conditioned Coproduction (Dependent Origination), 86, 93
Confucianism, 94-95
Conze, E., 86
Copernican Principle, 67
Creation myths, 104, 136; Canaanite (Ugarit), 111; Enuma Elish, 111
Cullman, O., 111-12

Daniel, 114-16, 139
Davisson and Germer, 28
deBroglie, L., 27-28, 31, 33
Denbigh, K., 2, 166-67, 174
Determinism, 24, 40, 53
Dharma, 81
Diffraction, 30
Dirac, P.A.M., 65

INDEX

Dodd, C. H., 113
Doppler effect, 56; blue shift, 66; red shift, 56
Dunne, J. W., 169
Dyson, F. J., 65-66, 140

Eddington, Sir A. 6, 48, 150, 174
Eight-Fold Path, 85-87
Einstein, A., 10-14, 20-23, 25, 62; formula $E = mc^2$, 19, 38, 63, 159; Nobel prize, 24, 26; photoelectric effect, 26
Electron volt, 59, 63; definition, 59; GeV, definition, 59
Elementary particle physics, 59-61; definition, 55
Eliade, M., 1, 103-5, 108-9, 144, 172; sacred and profane time, 103, 165
Eliot, C., 160
Eliot, T. S., 171
Entropy, 44-49, 52-53
Equivalence, Principle of, 20
Eschatology, 113, 153-54
Ether, 11-13
Exodus, 105-6

Four Noble Truths, 85
Fraser, J. T., 22-23, 149, 170-71, 174

Gamow, G., 58
Genesis, 56, 110-12, 136-38, 149
Govinda, Lama A., 92, 148, 157, 163
Grand Unified Theories, GUT, 61, 64, 71, 145; Super-GUT or Supergravity, 61, 170
Gravitation, 20-21, 61

Hawking, S. W., 65; Collins and, 68
Heisenberg, W., 29, 31, 36; matrix mechanics, 31; relation to Black Holes, 65; relation to complementarity, 33; Uncertainty Principle, 29-31, 38, 47-48, 70, 147, 159
Hinayana, 88
Hinduism, 75-79, 81, 171; cosmology, 79-82, 134-35, 146-47; view of time, 82-84
Husserl, E., 151, 154

I Ching (Classic of changes), 96, 155; hexagrams, 96
I-Hsing, 97, 142
Isaac, 105, 108
Israelites, 105-10, 116
Isvara, 80
I-Thou. *See* Buber, M.

Jainism, 77
Joshua, 106
Jastrow, R., 66

Kairos, 107
Kalpa, 81, 89
Karma, 78, 81
K-meson, 50, 54
Krishna, 77-78
Kronos, 6

Lao-tzu, 97-98, 135
Layzer, D., 49
Lee, T. D., 39, 161-62
Light, 21, 26-27, 149; cone, 17-19; velocity of, 11-17, 56
Long, C. H., 136
Longfellow. W. W., 172

Macroscopic world, 25, 44
Madhva, 79, 82
Madhyamikas, 88
Magnetic monopoles, 58-59
Mahabharata, 77-78
Mahayana, 88-93
Marsh, J., 112-13

INDEX

Mass, 19, 159
Maya, 80-81
Mayan calendar, 142
Mercury, precession of, 21
Merleau-Ponty, M., 151, 154
Messiah, 110, 112, 153
Messianic Age, 112, 114, 154
Michelson and Morley, 12-13
Microscopic world, 25, 44
Minkowski, H., 14-15
Moksha, 81, 140, 144
Moses, 105-6

Nagarjuna, 88
Nagasena, 90
Needham, J., 96, 100
New Testament, 114, 135, 139-40
Newton, Isaac, 2, 10
Nirvana, 86-87, 89-92, 140
Nucleus, size of, 25
Nyaya school, 79, 82

Old Testament, 108-15, 136-40

Pannikar, R., 144, 174
Park, D., 2, 43, 166-67, 170, 172
Penzias and Wilson, 57, 60
Philosophy, of becoming, 22, 61-62; of being, 22, 61-62
Photoelecric effect. See Einstein, A.
Planck, M., 26, 28, 30-31
Planck's constant, 26
Plato, 144-45
Prakriti, 80
Prigogine, I., 51-53, 54
Pudgalavadins, 88
Puranas, 78, 80
Pureland sect, 89, 95
Purva-Mimamsa, 79, 82

"Quantum foam." See Wheeler, J. A.
Quantum Mechanics, 31-33, 34

Quantum Theory, 24, 34-35; Observer Theory, 35-37

Ramanuja, 79, 82
Ramayana, 78
Red Giant, 64, 139
Relativity, Theory of, 12, 158-9; general, 20-23; Principle of, 11; special, 12-19; transformation equations, 14-15, 159
Revelation, 114-16, 139-40
Riemann, G. F. D., 20
Robinson, R. H., 85

St. Augustine, 56, 113, 117-20, 136-38, 145, 174; concept of time, 2, 118-20, 130, 162; God's "eternal now," 148
Sakharov, A., 49, 142, 170
Samsara, 77, 86, 92
Sankhya school, 79
Sarvastivadin, 88
Schlegel, R., 48, 147
Schramm, D. N., 62, 72
Schroedinger, E., 31-32, 46, 148
Shankara, 79, 82-83
Shiva, 80
Simultaneity, 16-17
Skandhas, 87-88
Space, contraction, 16, 19; curvature, 20; Interdependence with time, 14-16; quantization, 57-59, 161-62
"Spectrum of Intellectual Pursuits," 5, 74
Spinoza, 24
Sutras, 79
Suzuki, D. T., 91, 157, 161
Supergravity. See Grand Unified Theories

Tantraism, 91
Tao te Ching, 98-100, 135
Taoism, 95-100, 171; cosmology,

99; Tao, 98-100, 137, 142, 146-47, 171
Teilhard de Chardin, 117-18, 127-30, 140-41, 152-53, 155; Omega Point, 129-31, 141, 153
Thermodynamics, 44; First Law of, 46; Second Law of, 46-48
Time, arrow of, 48-51, 150-56; biblical, 107-12, 153-55; cosmologic arrow, 50-51, 53, 57; cyclic, 49, 67, 102-5, 142, 146-47; dilation, 16, 19; historical arrow, 49-51, 53; interdependence with space, 14-16, 157-60; quantization, 37-39, 161-62; reversible, 42-44, 51-52, 147-48; thermodynamic arrow, 48-51, 53; timelessness, 83-84, 144-49
Transformation equation, See Relativity.
Trefil, J., 61, 137
Twin paradox, 21

Uncertainty Principle. See Heisenberg, W.
Universe, age of, 47, 59, 72; cyclic, 49, 67, 142, 146-47; inflationary, 59; open or closed, 62-67
Upanishads, 76, 80; *Chandogya,* 81; *Svetasvatara,* 77, 80

Vaisesika school, 79, 82

Vedanta, 79, 82-83
Vedas, 75; *Atharva,* 76, 83-84; *Rig,* 75, 81
Vishnu, 80
Von Franz, M., 96
Von Rad, G., 108
Von Schiller, 177

Wave-particle duality, 27-29, 168
Welch, H., 100
Wheeler, J. A., 36-39, 63, 68-72, 136-37, 145; concept of time, 70, 147, 162-63; Law of Mutability, 70, 136; matter and space, 21, 57; "quantum foam," 38, 62, 142, 147, 170
Whitehead, A. N., 117, 120-24, 130, 140-41, 172; Concept of time, 123-24, 152-53, 155, 157-60, 162; religious views, 122-23
Wu Ching (5 Classics), 95
Wu-wei, 98

Yahweh, 105-9, 153
Yin-yang, 96, 167-68; symbol, 97
Yoga, Buddhist, 90, 148; Hindu school, 79; yogin (practitioner), 90
Yogacara, 88
Yuga, 80-81

Zurvan, 5

About the Author

Lawrence Fagg is Research Professor in Nuclear Physics at the Catholic University of America in Washington, D.C. and a Fellow of the American Physical Society. In addition to a Ph.D. in Nuclear Physics from John Hopkins University, he holds a Master's degree in Religion from George Washington University.

A vice president of the Institute on Religion in an Age of Science and a member of the International Society for the Study of Time, Dr. Fagg has lectured and presented seminars in the area of science and religion for the Forum for the Humanities of the Washington School of Psychiatry and also for the Institute on Religion in an Age of Science. He has lectured on the subject of time in physics and religion at meetings of the George Mason University and the Catholic University of America. He has also lectured on his work in physics, which concerns electron scattering from nuclei, at universities, laboratories, and conferences in the United States, Canada, Europe, Japan, and Australia.

QUEST BOOKS
are published by
The Theosophical Society in America,
a branch of a world organization
dedicated to the promotion of brotherhood and
the encouragement of the study of religion,
philosophy, and science, to the end that man may
better understand himself and his place in
the universe. The Society stands for complete
freedom of individual search and belief.
In the Theosophical Classics Series
well-known occult works are made
available in popular editions.

Additional Quest books are available on a wide spectrum of subject matter such as transpersonal psychology, philosophy, comparative religion, occultism, meditation, etc. Write for our free catalog.

The Theosophical Publishing House
306 West Geneva Road
Wheaton, Illinois 60189